ATTACKING
TRIGONOMETRY
PROBLEMS

David S. Kahn

DOVER PUBLICATIONS, INC.
Mineola, New York

Copyright

Copyright © 2015 by David S. Kahn
All rights reserved.

Bibliographical Note

Attacking Trigonometry Problems is a new work,
first published by Dover Publications, Inc., in 2015.

International Standard Book Number

ISBN-13: 978-0-486-78967-5
ISBN-10: 0-486-78967-5

Manufactured in the United States by Courier Corporation
78967501   2015
www.doverpublications.com

To the Reader:

Welcome to trigonometry! If you are like most students, this is the last mathematics course that you will have to take before calculus, or in some cases, ever! One of the difficulties of trigonometry is that you will have to combine several of the skills that you have developed in previous classes, namely, geometry, algebra, and graphing. Furthermore, you will have to use both of these in algebraic problems and in word problems. This may sound intimidating, but it doesn't have to be. Trigonometry can be challenging, but that doesn't mean that you can't handle it. Thanks to the book that you are about to use, you will be able to attack trigonometry and successfully conquer it.

One of the keys to doing well in trigonometry is memorizing. We recognize that many of you are not excited about the idea of memorizing the various ratios and formulas, but you will find that you will do much better, and have an easier time with trig, if you do. We will point out in the text what you need to memorize. We suggest that you do so!

Another key is to drill. This goes along with memorization. It is important to get comfortable with using the trig ratios in a variety of ways, and it will be much easier if you do all of the practice problems to reinforce the concepts. If you do so, you will find that trig isn't so hard after all.

We have organized this book so that you can proceed from one topic of trig to another, or you can jump to the topics that you want to work on. The book is divided into 14 units, each of which will teach you what you need to know to do well in that topic. This is not designed to be an exhaustive treatise on trigonometry, nor is it designed to be a textbook. Rather, this book focuses on the essentials, and how to master the problems. We suggest that you read through each unit completely, do all of the exercises, and complete all of the practice problems. Each example and problem has a complete explanation to help you understand how to solve the problem correctly. There are many good textbooks on trigonometry, and after you have worked through a unit, you may want to refer to a textbook for further practice.

Trigonometry can be a fun and useful area of mathematics. After you have gone through this book, you will be able to handle the trig on your exams with ease, and you will be prepared for calculus. Are you ready? Then it's time to Attack Trigonometry!

Acknowledgments

First of all, I would like to thank Nicole Maisonet for her excellent drawings. It is a tedious task and she did it with grace and enthusiasm. Next, I would like to thank Magan Farraj for working through all of the problems, double-checking my calculations. I owe a lifetime debt to my father, Peter Kahn, and to my dear friend, Arnold Feingold, who encouraged my interest in Mathematics and have always been there to guide me through the rough spots. And finally, I would like to thank the very many students whom I have taught and tutored, who have never hesitated to correct me when I am wrong, and who provide the fulfillment that I so deeply derive from teaching math.

Table of Contents

Unit One	The Basic Trig Ratios	1
Unit Two	Special Triangles	10
Unit Three	Trig Ratios for Other Angles	13
Unit Four	Degrees and Radians	27
Unit Five	The Reciprocal Functions	33
Unit Six	Some Basic Trigonometry Problems	42
Unit Seven	Sine and Cosine Graphs	70
Unit Eight	Graphing Tangent, Cotangent, Secant, and Cosecant	90
Unit Nine	Inverse Trigonometric Functions	102
Unit Ten	Basic Trigonometric Identities and Equations	113
Unit Eleven	More Trigonometric Identities	121
Unit Twelve	Trigonometric Angle Formulas	129
Unit Thirteen	The Law of Sines	142
Unit Fourteen	The Law of Cosines and Area Formulas	155

Attacking Trigonometry Problems

UNIT ONE

The Basic Trig Ratios

Trigonometry consists of learning how to use six different functions, or ratios, which show up in a surprisingly large number of places. Where do they come from? A good place to start is with some basic geometry. Remember similar triangles? If two triangles are similar, then they have equal angles and the ratio of their sides is the same. For example,

Figure 1

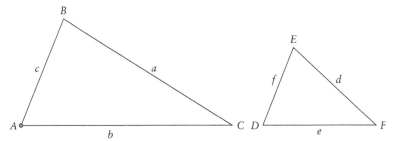

If the two triangles above are similar, then $\angle A = \angle D$, $\angle B = \angle E$, and $\angle C = \angle F$, and $\frac{a}{d} = \frac{b}{e} = \frac{c}{f}$.

Let's look at two similar right triangles:

Figure 2

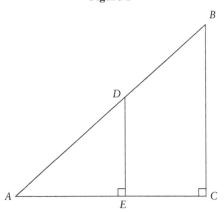

Notice that $\triangle ABC$ is similar to $\triangle ADE$ because each contains a right angle (C & E) and the same angle (A), and so the third angle must also be the same (because the measures

of the angles in a triangle add to 180°). This means that $\frac{AD}{AB} = \frac{AE}{AC} = \frac{DE}{BC}$. If we had the following set of right triangles, the corresponding ratios would all be equal.

Figure 3

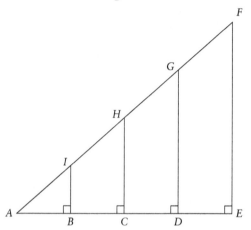

In fact, for *any* right triangle that has an angle with a measure equal to the measure of angle *A*, the ratios are the same as those of *any other* right triangle that has an angle with measure equal to that of *A*. This is *the* essential fact of Trigonometry and can be used in many powerful ways.

For the triangle below, these are the three basic trigonometric ratios to learn:

Figure 4

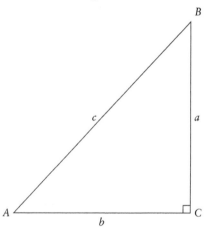

UNIT ONE: The Basic Trig Ratios

The *sine* of angle A is the ratio $\frac{a}{c}$.

The *cosine* of angle A is the ratio $\frac{b}{c}$.

The *tangent* of angle A is the ratio $\frac{a}{b}$.

We usually abbreviate sine as sin, cosine as *cos*, and tangent as *tan*, and write these ratios using the following notation:

$$\sin A = \frac{a}{c}, \quad \cos A = \frac{b}{c}, \quad \tan A = \frac{a}{b}.$$

You will want to get comfortable with these ratios. There is an easy way to memorize them.

Figure 5

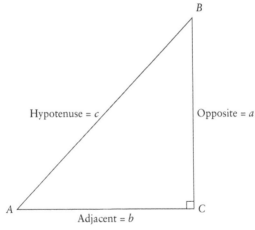

There are three sides from the perspective of angle A: a is the side opposite angle A, b is the side adjacent to angle A, and c is the hypotenuse. Therefore, we can think of:

$$\sin A \text{ as } \frac{opposite}{hypotenuse}$$

$$\cos A \text{ as } \frac{adjacent}{hypotenuse}$$

$$\tan A \text{ as } \frac{opposite}{adjacent}.$$

This gives the traditional mnemonic:

SOH CAH TOA

which stands for: $\sin = \dfrac{opposite}{hypotenuse}$, $\cos = \dfrac{adjacent}{hypotenuse}$, $\tan = \dfrac{opposite}{adjacent}$.

Example 1:

Figure 6

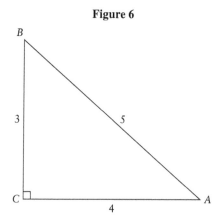

In the triangle above, sin A is the side opposite angle A (3) divided by the hypotenuse (5), so $\sin A = \dfrac{3}{5}$. Similarly, cos A is the side adjacent to angle A (4) divided by the hypotenuse (5), so $\cos A = \dfrac{4}{5}$. What is tan A? $\tan A = \dfrac{3}{4}$. Got the idea? Of course, we could also find the trig ratios of the other acute angle, B. Now, sin B is the side opposite angle B (4) divided by the hypotenuse (5), so $\sin B = \dfrac{4}{5}$. Similarly, cos B is the side adjacent to angle B (3) divided by the hypotenuse (5), so $\cos B = \dfrac{3}{5}$ and $\tan B = \dfrac{4}{3}$.

UNIT ONE: The Basic Trig Ratios

Example 2:

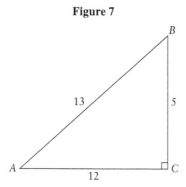

Figure 7

What are the three trig ratios of angle A?
$\sin A = \dfrac{5}{13}$, $\cos A = \dfrac{12}{13}$, and $\tan A = \dfrac{5}{12}$
What if we asked for the three trig ratios of angle B?
They are $\sin B = \dfrac{12}{13}$, $\cos B = \dfrac{5}{13}$, and $\tan B = \dfrac{12}{5}$.

Notice how in these examples, $\sin A = \cos B$ and $\cos A = \sin B$? This is not a coincidence! Remember that the sum of the two acute angles in a right triangle is $90°$. This means that angle $A = (90° - B)$ and angle $B = (90° - A)$. Therefore, in any right triangle, $\sin A = \cos(90° - A)$ and $\cos A = \sin(90° - A)$.

Example 3:

Figure 8

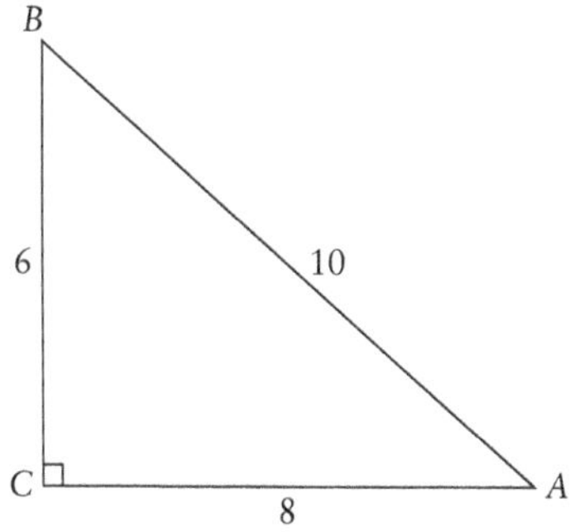

For this triangle, $\sin A = \dfrac{6}{10} = \cos B$, and $\sin B = \dfrac{8}{10} = \cos A$.

By the way, the tangent of one of the acute angles is the reciprocal of the tangent of the other acute angle. In other words, $\tan A = \dfrac{1}{\tan(90° - A)}$ (and vice versa).

Notice that in Figure 8, $\tan A = \dfrac{6}{8}$, and $\tan B = \dfrac{8}{6}$.

Notice that we have not been finding the trig ratios for the right angle. Right now, we only know how to find the trig ratios for an angle between $0°$ and $90°$. Later, we will learn how to find the trig ratios for an angle of any measure.

Time to practice!

Practice Problems

Practice problem 1:

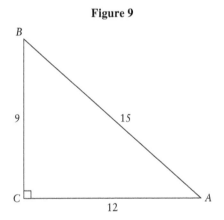

Figure 9

Find: sin A _____   sin B _____
        cos A _____   cos B _____
        tan A _____   tan B _____

UNIT ONE: The Basic Trig Ratios

Practice problem 2:

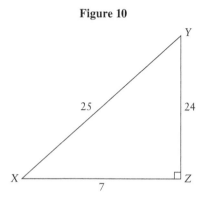

Figure 10

Find:
    sin X _____   sin Y _____
    cos X _____   cos Y _____
    tan X _____   tan Y _____

Practice problem 3:

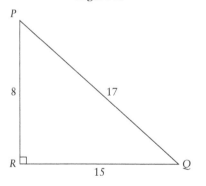

Figure 11

Find:
    sin P _____   sin Q _____
    cos P _____   cos Q _____
    tan P _____   tan Q _____

Practice problem 4:
If $\sin(2x-5) = \cos(5x+25)$, find x.

Practice problem 5:
If $\cos(x^2 - 20) = \sin(12 + x^2)$, find x, where the angle is in degrees.

Solutions to the Practice Problems

Solution to practice problem 1:

Remember our mnemonic SOH CAH TOA! Here, the side opposite angle A is 9, the side adjacent to angle A is 12, and the hypotenuse is 15. Therefore, we get:

$$\sin A = \frac{9}{15} = \frac{3}{5}, \quad \cos A = \frac{12}{15} = \frac{4}{5}, \quad \tan A = \frac{9}{12} = \frac{3}{4}.$$

Similarly, the side opposite angle B is 12, the side adjacent to angle B is 9, and the hypotenuse is 15. We get:

$$\sin B = \frac{12}{15} = \frac{4}{5}, \quad \cos B = \frac{9}{15} = \frac{3}{5}, \quad \tan B = \frac{12}{9} = \frac{4}{3}.$$

Solution to practice problem 2:

Here, the side opposite angle x is 24, the side adjacent to angle x is 7, and the hypotenuse is 25. Therefore, we get:

$$\sin y = \frac{24}{25}, \quad \cos y = \frac{7}{25}, \quad \tan y = \frac{24}{7}.$$

Now, the side opposite angle y is 7, the side adjacent to angle y is 24, and the hypotenuse is 25. We get:

$$\sin y = \frac{7}{25}, \quad \cos y = \frac{24}{25}, \quad \tan y = \frac{7}{24}.$$

Solution to practice problem 3:

Here, the side opposite angle P is 15, the side adjacent to angle P is 8, and the hypotenuse is 17. Therefore, we get:

$$\sin P = \frac{15}{17}, \quad \cos P = \frac{8}{17}, \quad \tan P = \frac{15}{8}.$$

Now, the side opposite angle Q is 8, the side adjacent to angle Q is 15, and the hypotenuse is 17. We get:

$$\sin Q = \frac{8}{17}, \quad \cos Q = \frac{15}{17}, \quad \tan Q = \frac{8}{15}.$$

UNIT ONE: The Basic Trig Ratios

Solution to practice problem 4:

Remember our rule that $\sin A = \cos(90° - A)$. This means that we can rewrite the equation:

$$\sin(2x-5) = \cos(5x+25)$$
$$\sin(2x-5) = \sin[90-(5x+25)]$$

If the two sines are equal, then the angles must be equal. This gives us:

$$(2x-5) = [90-(5x+25)].$$

Now we can use some simple algebra!

$$2x - 5 = 90 - 5x - 25$$
$$2x - 5 = 65 - 5x$$
$$7x = 70$$
$$x = 10$$

Solution to practice problem 5:

Again, we use our rule that $\sin A = \cos(90° - A)$. We can rewrite the equation:

$$\cos(x^2 - 20) = \sin(12 + x^2)$$
$$\cos(x^2 - 20) = \cos[90 - (12 + x^2)]$$

If the two cosines are equal, then the angles must be equal. This gives us:

$$(x^2 - 20) = [90 - (12 + x^2)].$$

Algebra time!

$$x^2 - 20 = 90 - 12 - x^2$$
$$x^2 - 20 = 78 - x^2$$
$$2x^2 = 98$$
$$x^2 = 49$$
$$x = \pm 7$$

UNIT TWO

Special Triangles

Now that we have learned the three basic trig ratios, let's learn how to find the sine, cosine, and tangent of angles in *special triangles*.

You should remember from Geometry that an equilateral triangle has some special properties. First, the measures of all three angles are 60°. Second, the lengths of all the sides are equal. If we drop a perpendicular line from a vertex of an equilateral triangle to the opposite side, the triangle is cut into two congruent triangles. In the figure below, we can see that the measures of the angles of triangle ABD and triangle CBD are then $30° - 60° - 90°$:

Figure 1

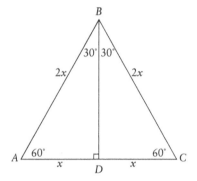

Next, if the sides of the triangle each have length $2x$, then side AD has length x, and we can use the Pythagorean Theorem to find that side BD has length $x\sqrt{3}$:

Figure 2

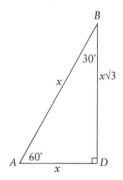

UNIT TWO: Special Triangles

Now we can find the trig functions for angles of measures 30° and 60°:

$$\sin 30° = \frac{x}{2x} = \frac{1}{2}; \quad \cos 30° = \frac{x\sqrt{3}}{2x} = \frac{\sqrt{3}}{2}; \quad \tan 30° = \frac{x}{x\sqrt{3}} = \frac{1}{\sqrt{3}}$$

$$\sin 60° = \frac{x\sqrt{3}}{2x} = \frac{\sqrt{3}}{2}; \quad \cos 60° = \frac{x}{2x} = \frac{1}{2}; \quad \tan 60° = \frac{x\sqrt{3}}{x} = \sqrt{3}.$$

You should try to commit these to memory because they show up often in many types of math problems. In a little bit, we will learn an easy way to memorize these.

Now, let's look at another special triangle. A square has some special properties as well. First, all of the sides are congruent. Second, all of the angles are right angles. If you take a square of side x and draw a diagonal, you get two isosceles right triangles.

Figure 3

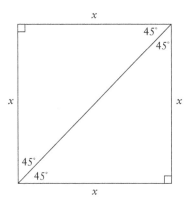

We can use the Pythagorean Theorem to find the length of the diagonal, which is $x\sqrt{2}$. This triangle is often referred to as a 45°–45°–90° triangle. Now we can find the trig ratios of a 45°-angle:

$$\sin 45° = \frac{x}{x\sqrt{2}} = \frac{1}{\sqrt{2}}; \quad \cos 45° = \frac{x}{x\sqrt{2}} = \frac{1}{\sqrt{2}}; \quad \tan 45° = \frac{x}{x} = 1.$$

Note that $\sin 45° = \cos 45°$. Why should this make sense?

By the way, many people rationalize the denominator of $\frac{1}{\sqrt{2}}$ by multiplying the numerator and denominator by $\sqrt{2}$: $\frac{1}{\sqrt{2}}\left(\frac{\sqrt{2}}{\sqrt{2}}\right) = \frac{\sqrt{2}}{2}$. This makes memorizing the sines and cosines of the special angles a little easier.

Let's make a table of the trig functions that we just learned:

	30°	45°	60°
sin	$\dfrac{1}{2}$	$\dfrac{\sqrt{2}}{2}$	$\dfrac{\sqrt{3}}{2}$
cos	$\dfrac{\sqrt{3}}{2}$	$\dfrac{\sqrt{2}}{2}$	$\dfrac{1}{2}$
tan	$\dfrac{1}{\sqrt{3}}$	1	$\sqrt{3}$

The reason that these are called the *special* angles is that we can find the exact values of the trig functions for these angles. For almost all other angles, however, we approximate the trig ratios with decimal values.

UNIT THREE
Trig Ratios for Other Angles

Now let's learn how to find the trig values for other angles. Suppose that you draw a circle of radius 1 (the "unit circle"), centered at the origin. Pick a point in Quadrant I on the circle and draw the radius from the origin to that point.

Figure 1

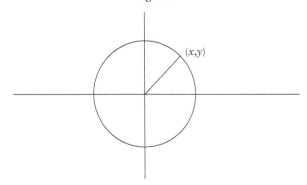

We can construct a right triangle using that radius. The lengths of the legs are equal to the coordinates x and y, respectively, and the hypotenuse of the right triangle is 1. If we call the angle between the positive x-axis and the radius θ, then we can see that $\sin\theta = \frac{y}{1}$, $\cos\theta = \frac{x}{1}$, and $\tan\theta = \frac{y}{x}$. In other words, the coordinates of the point (x, y) are $(\cos\theta, \sin\theta)$.

Figure 2

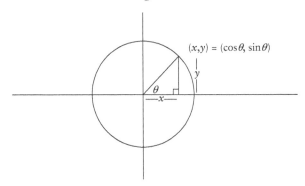

Why is this useful? We can now use this information to find the sine, cosine, and tangent of any angle.

Example 1: Let's find the sine, cosine, and tangent of 0°.
These ratios are simply the coordinates when the angle between the radius and the positive *x*-axis is $\theta = 0°$. The coordinates on the unit circle are $(1, 0)$:

Figure 3

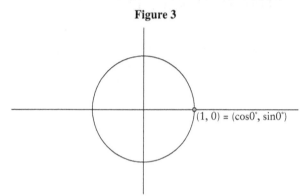

Thus, $\sin 0° = 0$, $\cos 0° = 1$, and $\tan 0° = \frac{0}{1} = 0$.

Example 2: Let's do it again for $\theta = 90°$. Here, the coordinates for the angle between the radius and the positive *x*-axis are $(0, 1)$:

Figure 4

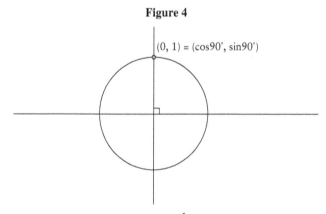

Thus, $\sin 90° = 1$, $\cos 90° = 0$, and $\tan 90° = \frac{1}{0} = undefined$. Notice that the tangent is undefined at 90°. You will find that the trig functions other than sine and cosine are undefined at certain angles. These will occur with trig functions that are formed with the sine and cosine functions in their denominators. They will be undefined

UNIT THREE: Trig Ratios for Other Angles   15

at those angles where the sine or cosine is zero because you can't have zero in the denominator of a fraction. You will learn more about that in the next unit.

Example 3: Let's keep going to 180°:

Figure 5

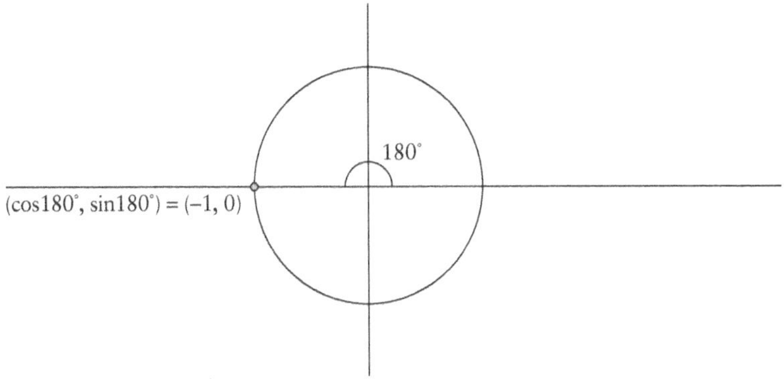

$(\cos 180°, \sin 180°) = (-1, 0)$

180°

Here, the coordinates of the intersection are $(-1, 0)$. So, $\sin 180° = 0$, $\cos 180° = -1$, and $\tan 180° = \dfrac{0}{-1} = 0$.

Example 4: Finally, let's find the ratios for 270°:

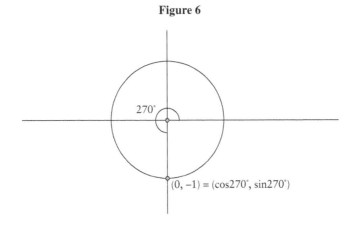

Figure 6

270°

$(0, -1) = (\cos 270°, \sin 270°)$

Here, the coordinates of the intersection are (0, −1). So, $\sin 270° = -1$, $\cos 270° = 0$, and $\tan 270° = \frac{-1}{0} = undefined$.

You should memorize the sine, cosine, and tangent for 0°, 90°, 180°, and 270°. We can make a table to help us memorize these values.

	0°	90°	180°	270°
Sin	0	1	0	−1
Cos	1	0	−1	0
Tan	0	undefined	0	undefined

Example 5: Let's find the trig values for $\theta = 150°$. If we draw the unit circle, we are looking for the coordinates of the point where the radius intersects the circle when the angle between the radius and the positive x-axis is 150°:

Figure 7

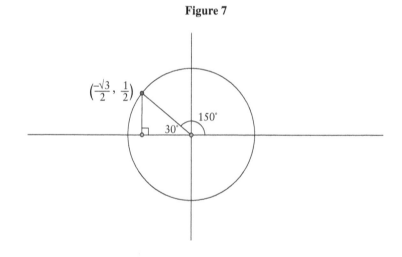

But however, notice that the angle between the radius and the *negative* x-axis is 30°. This is called the *reference angle* and we can use it, rather than 150°, to find the trig ratios. We can use the reference angle to make a 30°–60°–90° triangle with the negative x-axis. The coordinates of the point of intersection are $\left(-\frac{\sqrt{3}}{2}, \frac{1}{2}\right)$. This means that $\sin 150° = \frac{1}{2}$, $\cos 150° = -\frac{\sqrt{3}}{2}$, and $\tan 150° = -\frac{1}{\sqrt{3}}$. Notice that these are the same values as $\sin 30°$, $\cos 30°$, and $\tan 30°$, except that cosine and tangent are negative.

UNIT THREE: Trig Ratios for Other Angles 17

It is important to understand what the reference angle is. When we are looking for the trig ratio of any angle other than an acute one, we find it by using the unit circle. We then draw the radius that is the measure of the angle we are evaluating. The radius will always make an acute angle with either the positive or negative *x*-axis. That angle is the reference angle, and we use the trig values of that acute angle. Remember that the reference angle is always an acute angle, and it is always formed between the radius and the *x*-axis, *never* the *y*-axis.

Example 6: Now, let's find sine, cosine, and tangent of 225°.

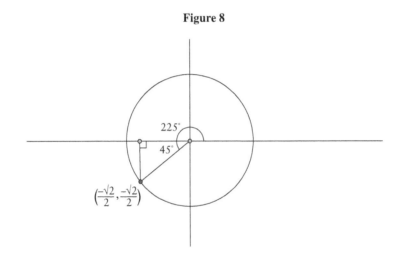

Figure 8

Here, we use the angle between the radius and the negative *x*-axis to find the reference angle of $225° - 180° = 45°$. We can use the reference angle to make a $45°-45°-90°$ triangle with the *x*-axis. This time, the coordinates are $\left(-\frac{\sqrt{2}}{2}, -\frac{\sqrt{2}}{2}\right)$. (Why?) Therefore, $\sin 225° = -\frac{\sqrt{2}}{2}$, $\cos 225° = -\frac{\sqrt{2}}{2}$, and $\tan 225° = 1$. Again, these are the same values as sin 45°, cos 45°, and tan 45°, except that the sine and cosine are negative.

Why do the signs change? In Quadrant I, the *x*- and *y*-coordinates of a point will be both positive. However, in Quadrant II the *x*-coordinate will be negative while the *y*-coordinate stays positive. Because cosine is equal to the value of the *x*-coordinate, cosine is negative, while sine stays positive. Furthermore, because tangent is $\frac{y}{x}$, and one coordinate is negative, tangent is negative. In Quadrant III, both coordinates are negative, so sine and cosine are both negative. The tangent is positive because it is the ratio of two negative numbers. Finally, in Quadrant IV, the *x*-coordinate is positive and the *y*-coordinate is negative. So, sine is negative, cosine is positive, and tangent is negative.

There is an easy way to remember the signs of the trig ratios in the various quadrants. Look at the figure below:

Figure 9

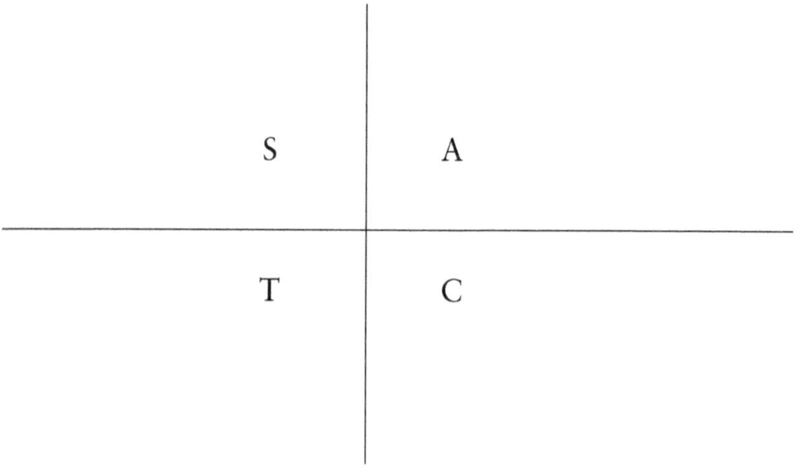

S	A
T	C

The mnemonic is.

All **S**tudents **T**ry **C**andy

which stands for:

All trig ratios are positive in Quadrant I.
Sine is positive in Quadrant II (and the others are negative).
Tangent is positive in Quadrant III (and the others are negative).
Cosine is positive in Quadrant IV (and the others are negative).

Let's do another, just to make sure that you get the idea.

Example 7: Let's find sine, cosine, and tangent of $300°$. Notice that we stopped drawing the unit circle. The circle isn't really necessary because the acute angle that the radius makes with the *x*-axis is what we are interested in.

Figure 10

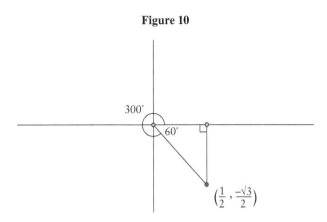

Here, we use the angle between the radius and the positive x-axis. This is the reference angle and has a measure of $360° - 300° = 60°$. We can use the reference angle to make a $30°-60°-90°$-triangle with the x-axis. This time, the coordinates are $\left(\frac{1}{2}, -\frac{\sqrt{3}}{2}\right)$. Therefore, $\sin 300° = -\frac{\sqrt{3}}{2}$, $\cos 300° = \frac{1}{2}$, and $\tan 300° = -\sqrt{3}$. These are the same values as sin 60°, cos 60°, and tan 60°, except that the sine and tangent are negative, which we expected from our mnemonic.

What if we used an angle greater than 360°? Well, we would just go around the axes in full circles as many times as needed and then around in part of a circle until we get to our destination.

Example 8: Suppose we wanted to find sine, cosine, and tangent of 420°. We would go once around the x-axis, and another 60°.

Figure 11

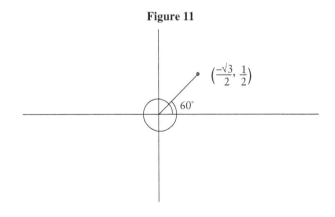

Thus, the sine, cosine, and tangent of 420° are the same as for 60°, namely, $\sin 60° = \frac{\sqrt{3}}{2}$, $\cos 60° = \frac{1}{2}$, and $\tan 60° = \sqrt{3}$. They are all positive because we are back in Quadrant I.

Finally, what about if we measure the angle going clockwise instead of counterclockwise? We call this a *negative angle*.

Example 9: Let's find the sine, cosine, and tangent of −210°. Now we find the angle by going around the axes in the opposite direction:

Figure 12

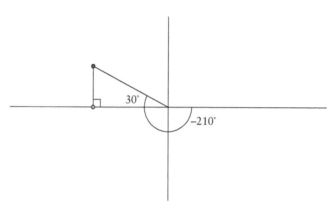

Thus, going to −210° is the same as going to 150° because −210° + 360° = 150°. Thus, the trig ratios are $\sin(-210°) = \frac{1}{2}$, $\cos(-210°) = -\frac{\sqrt{3}}{2}$, and $\tan(-210°) = -\frac{1}{\sqrt{3}}$.

Let's do some practice problems.

Practice Problems

Practice problem 1: Find the sine, cosine, and tangent of 120°.
Practice problem 2: Find the sine, cosine, and tangent of 135°.
Practice problem 3: Find the sine, cosine, and tangent of 210°.
Practice problem 4: Find the sine, cosine, and tangent of 240°.
Practice problem 5: Find the sine, cosine, and tangent of 315°.
Practice problem 6: Find the sine, cosine, and tangent of 330°.
Practice problem 7: Find the sine, cosine, and tangent of 480°.
Practice problem 8: Find the sine, cosine, and tangent of 870°.
Practice problem 9: Find the sine, cosine, and tangent of −45°.
Practice problem 10: Find the sine, cosine, and tangent of −300°.

Solutions to the Practice Problems

Solution to practice problem 1: *Find the sine, cosine, and tangent of* 120°.
First, let's draw a picture so that we can find the reference angle:

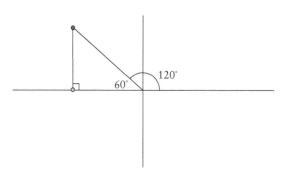

We can see that the acute angle with the *x*-axis is 60° (the reference angle), so we need to find the sine, cosine, and tangent of 60°, and then remember that the sine will be positive and the cosine and tangent will be negative because we are in Quadrant II. We get:

$$\sin 120° = \frac{\sqrt{3}}{2}, \quad \cos 120° = -\frac{1}{2}, \quad \text{and} \quad \tan 120° = -\sqrt{3}.$$

Solution to practice problem 2: *Find the sine, cosine, and tangent of* 135°.
First, let's draw a picture so that we can find the reference angle:

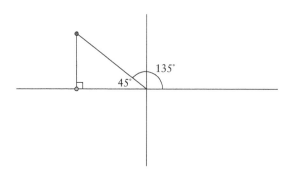

We can see that the acute angle with the *x*-axis is 45° (the reference angle), so we need to find the sine, cosine, and tangent of 45°. Remember that the sine will be positive and the cosine and tangent will be negative because we are in Quadrant II. We get:

$$\sin 135° = \frac{\sqrt{2}}{2}, \quad \cos 135° = -\frac{\sqrt{2}}{2}, \quad \text{and} \quad \tan 135° = -1.$$

Solution to practice problem 3: *Find the sine, cosine, and tangent of* 210°.
First, let's draw a picture so that we can find the reference angle:

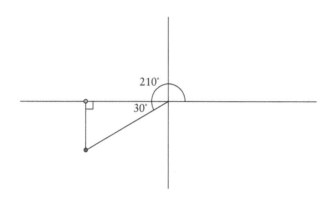

We can see that the acute angle with the *x*-axis is 30° (the reference angle), so we need to find the sine, cosine, and tangent of 30°. Remember that the tangent will be positive and the sine and cosine will be negative because we are in Quadrant III. We get:

$$\sin 210° = -\frac{1}{2}, \quad \cos 210° = -\frac{\sqrt{3}}{2}, \quad \text{and} \quad \tan 210° = \frac{1}{\sqrt{3}}.$$

Solution to practice problem 4: *Find the sine, cosine, and tangent of* 240°.
First, let's draw a picture so that we can find the reference angle:

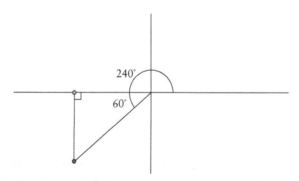

We can see that the acute angle with the *x*-axis is 60° (the reference angle), so we need to find the sine, cosine, and tangent of 60°. Remember that the tangent will be positive and the sine and cosine will be negative because we are in Quadrant III. We get:

$$\sin 240° = -\frac{\sqrt{3}}{2}, \quad \cos 240° = -\frac{1}{2}, \quad \text{and} \quad \tan 240° = \sqrt{3}.$$

Solution to practice problem 5: *Find the sine, cosine, and tangent of 315°.*
First, let's draw a picture so that we can find the reference angle:

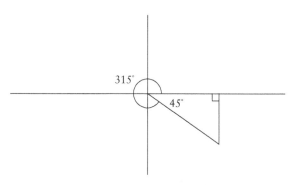

We can see that the acute angle with the *x*-axis is 45° (the reference angle), so we need to find the sine, cosine, and tangent of 45°. Remember that the cosine will be positive and the sine and tangent will be negative because we are in Quadrant IV. We get:

$$\sin 315° = -\frac{\sqrt{2}}{2}, \quad \cos 315° = \frac{\sqrt{2}}{2}, \quad \text{and} \quad \tan 315° = -1.$$

Solution to practice problem 6: *Find the sine, cosine, and tangent of 330°.*
First, let's draw a picture so that we can find the reference angle:

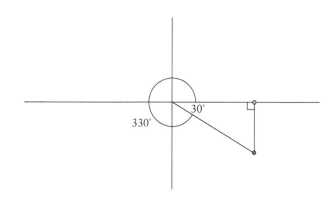

We can see that the acute angle with the x-axis is 30° (the reference angle), so we need to find the sine, cosine, and tangent of 30°. Remember that the cosine will be positive and the sine and tangent will be negative because we are in Quadrant IV. We get:

$$\sin 330° = -\frac{1}{2}, \quad \cos 330° = \frac{\sqrt{3}}{2}, \quad \text{and} \quad \tan 330° = -\frac{1}{\sqrt{3}}.$$

Solution to practice problem 7: *Find the sine, cosine, and tangent of 480°.*

First, let's draw a picture so that we can find the reference angle:

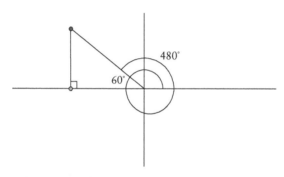

Here, we are going around the axes more than once (360°), and then continue another 120°, so we end up with the acute angle made with the x-axis as 60° (the reference angle). So we need to find the sine, cosine, and tangent of 60°. Remember that the sine will be positive and the cosine and tangent will be negative because we are in Quadrant II. We get:

$$\sin 480° = \frac{\sqrt{3}}{2}, \quad \cos 480° = -\frac{1}{2}, \quad \text{and} \quad \tan 480° = -\sqrt{3}.$$

Solution to practice problem 8: *Find the sine, cosine, and tangent of 870°.*

First, let's draw a picture so that we can find the reference angle:

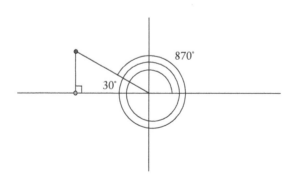

UNIT THREE: Trig Ratios for Other Angles 25

Here, we are going around the axes twice (2 : 360° = 720°) and then continue another 150°. We can see that the acute angle made with the *x*-axis is 30° (the reference angle), so we need to find the sine, cosine, and tangent of 30°. Remember that the sine will be positive and the cosine and tangent will be negative because we are in Quadrant II. We get:

$$\sin 870° = \frac{1}{2}, \quad \cos 870° = -\frac{\sqrt{3}}{2}, \quad \text{and} \quad \tan 870° = -\frac{1}{\sqrt{3}}.$$

Solution to practice problem 9: *Find the sine, cosine, and tangent of –45°.*

Here, we find the reference angle by going around the axes in the opposite direction:

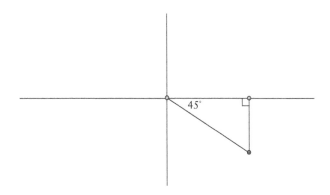

We can see that the acute angle with the *x*-axis is 45° (the reference angle), so we need to find the sine, cosine, and tangent of 45°. Remember that the cosine will be positive and the sine and tangent will be negative because we are in Quadrant IV. We get:

$$\sin(-45°) = -\frac{\sqrt{2}}{2}, \quad \cos(-45°) = \frac{\sqrt{2}}{2}, \quad \text{and} \quad \tan(-45°) = -1.$$

Solution to practice problem 10: *Find the sine, cosine, and tangent of –300°.*

Here, we find the reference angle by going around the axes in the opposite direction:

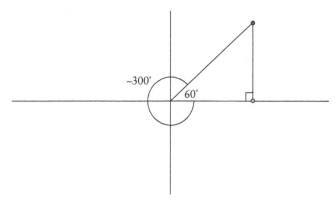

We can see that the acute angle made with the x-axis is 60° (the reference angle), so we need to find the sine, cosine, and tangent of 60°. Remember that the all of the values will be positive because we are in Quadrant I. We get:

$$\sin(-300°) = \frac{\sqrt{3}}{2}, \quad \cos(-300°) = \frac{1}{2}, \quad \text{and} \quad \tan(-300°) = \sqrt{3}.$$

UNIT FOUR
Degrees and Radians

When we measure the size of an angle, we usually express that size in degrees. Now we are going to learn another set of units to measure an angle. In the figure below, we have a circle of radius 1 ("the unit circle"). The circumference of this circle is $2\pi (1) = 2\pi$.

Figure 1

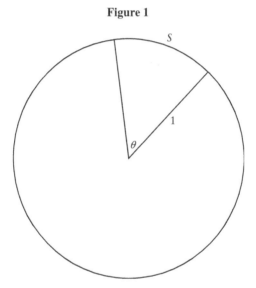

The central angle, θ, intercepts an arc of the circle that is directly related to the angle. We can find the length of the arc, s, easily. The ratio of the central angle θ to 360° is the same as the ratio of the arc s to the circumference. In other words,

$$\frac{\theta}{360°} = \frac{s}{2\pi(1)}.$$

We call the length of the arc the measure of the angle. Because the radians are found by taking the ratio of two lengths, they are dimensionless units. That is why we don't use a symbol for radians the way we do for degrees.

Here is how we convert an angle from degrees to radians.

Example 1: In the figure below, the central angle is 60°.

Figure 2

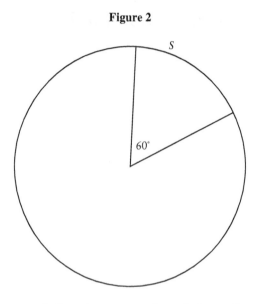

Now we can use our ratio formula to convert the angle into radians. We have

$$\frac{60°}{360°} = \frac{s}{2\pi}$$

$$\frac{1}{6} = \frac{s}{2\pi}$$

$$s = \frac{2\pi}{6} = \frac{\pi}{3}$$

So, $60° = \frac{\pi}{3}$ radians. By the way, we don't usually use a symbol for radians. If we are given the measure of an angle and it is not labeled in degrees, it is presumed to be in radians.

UNIT FOUR: Degrees and Radians

Example 2:

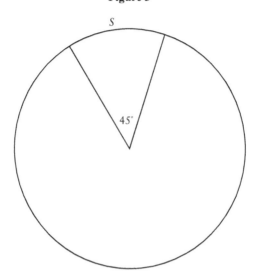

Figure 3

Here, the angle is 45°. We can convert this to radians:

$$\frac{45°}{360°} = \frac{s}{2\pi}$$

$$\frac{1}{8} = \frac{s}{2\pi}$$

$$s = \frac{2\pi}{8} = \frac{\pi}{4}$$

So $45° = \frac{\pi}{4}$.

We can simplify the ratios $\frac{\theta}{360°} = \frac{s}{2\pi}$ to $\frac{\theta}{180°} = \frac{s}{\pi}$. This means that $180° = \pi$ radians. Now we have an easy way to convert degrees to radians, or vice versa.

> To convert an angle from degrees to radians, multiply the angle measure by $\frac{\pi}{180}$.

> To convert an angle from radians to degrees, multiply the angle measure by $\frac{180}{\pi}$.

Let's do a few examples:

Example 3: Convert $30°$ to radians.

$$30\left(\frac{\pi}{180}\right) = \frac{\pi}{6}$$

Example 4: Convert $135°$ to radians.

$$135\left(\frac{\pi}{180}\right) = \frac{3\pi}{4}$$

Example 5: Convert $\frac{\pi}{2}$ to degrees.

$$\frac{\pi}{2}\left(\frac{180}{\pi}\right) = 90°$$

Example 6: Convert $\frac{11\pi}{6}$ to degrees.

$$\frac{11\pi}{6}\left(\frac{180}{\pi}\right) = 330°$$

In Calculus, you will work almost exclusively in radians. In Physics, you will usually see both used. Either way, you will need to be comfortable doing Trigonometry in both degrees and radians.

Time to practice!

UNIT FOUR: Degrees and Radians

Practice Problems

Practice Problem 1: Convert θ from degrees to radians: $\theta = 72°$.
Practice Problem 2: Convert θ from degrees to radians: $\theta = 225°$.
Practice Problem 3: Convert θ from degrees to radians: $\theta = 150°$.
Practice Problem 4: Convert θ from degrees to radians: $\theta = 540°$.
Practice Problem 5: Convert θ from degrees to radians: $\theta = 10°$.
Practice Problem 6: Convert θ from radians to degrees: $\dfrac{7\pi}{6}$.
Practice Problem 7: Convert θ from radians to degrees: $\dfrac{4\pi}{3}$.
Practice Problem 8: Convert θ from radians to degrees: $\dfrac{3\pi}{5}$.
Practice Problem 9: Convert θ from radians to degrees: 6π.
Practice Problem 10: Convert θ from radians to degrees: $\dfrac{\pi}{20}$.

Solutions to the Practice Problems

Solution to practice problem 1: *Convert θ from degrees to radians: $\theta = 72°$.*
Remember the rule! To convert an angle from degrees to radians, multiply the angle by $\dfrac{\pi}{180}$.
Here we get: $72\left(\dfrac{\pi}{180}\right) = \dfrac{2\pi}{5}$.

Solution to practice problem 2: *Convert θ from degrees to radians: $\theta = 225°$.*
Once again, to convert an angle from degrees to radians, multiply the angle by $\dfrac{\pi}{180}$.
Here we get: $225\left(\dfrac{\pi}{180}\right) = \dfrac{5\pi}{4}$.

Solution to practice problem 3: *Convert θ from degrees to radians: $\theta = 150°$.*
Multiply the angle by $\dfrac{\pi}{180}$: $150\left(\dfrac{\pi}{180}\right) = \dfrac{5\pi}{6}$.

Solution to practice problem 4: *Convert θ from degrees to radians: $\theta = 540°$.*
Multiply the angle by $\dfrac{\pi}{180}$: $540\left(\dfrac{\pi}{180}\right) = 3\pi$.

Solution to practice problem 5: *Convert θ from degrees to radians:* $\theta = 10°$.

Multiply the angle by $\dfrac{\pi}{180}$: $10\left(\dfrac{\pi}{180}\right) = \dfrac{\pi}{18}$.

Solution to practice problem 6: *Convert θ from radians to degrees:* $\dfrac{7\pi}{6}$.

Now we use the other rule! To convert an angle from radians to degrees, multiply the angle by $\dfrac{180}{\pi}$.

Here we get: $\dfrac{7\pi}{6}\left(\dfrac{180}{\pi}\right) = 210°$.

Solution to practice problem 7: *Convert θ from radians to degrees:* $\dfrac{4\pi}{3}$.

Once again, to convert an angle from radians to degrees, multiply the angle by $\dfrac{180}{\pi}$.
$\dfrac{4\pi}{3}\left(\dfrac{180}{\pi}\right) = 240°$

Solution to practice problem 8: *Convert θ from radians to degrees:* $\dfrac{3\pi}{5}$.

Multiply the angle by $\dfrac{180}{\pi}$: $\dfrac{3\pi}{5}\left(\dfrac{180}{\pi}\right) = 108°$.

Solution to practice problem 9: *Convert θ from radians to degrees:* 6π.

Multiply the angle by $\dfrac{180}{\pi}$: $6\pi\left(\dfrac{180}{\pi}\right) = 1080°$

Solution to practice problem 10: *Convert θ from radians to degrees:* $\dfrac{\pi}{20}$.

Multiply the angle by $\dfrac{180}{\pi}$: $\dfrac{\pi}{20}\left(\dfrac{180}{\pi}\right) = 9°$.

UNIT FIVE

The Reciprocal Functions

So far, we have been learning how to find the sine, cosine, and tangent of any angle. Now it is time to learn about three more trig functions, which are called the *reciprocal functions*. Each one of these functions is the reciprocal of one of the basic trig functions. These three functions are the *cosecant*, *secant*, and *cotangent*, which are abbreviated *csc*, *sec*, and *cot*, respectively. Here is how they are defined:

$$\csc\theta = \frac{1}{\sin\theta}$$
$$\sec\theta = \frac{1}{\cos\theta}$$
$$\cot\theta = \frac{1}{\tan\theta}.$$

For example, we know that $\sin 30° = \frac{1}{2}$, and so, $\csc 30° = \frac{1}{1/2} = 2$.

Similarly, $\sec 30° = \frac{1}{\cos 30°} = \frac{1}{\sqrt{3}/2} = \frac{2}{\sqrt{3}}$,

$\cot 30° = \frac{1}{\tan 30°} = \frac{1}{1/\sqrt{3}} = \sqrt{3}$.

Isn't that easy? These functions are valuable because, for example, you may do a problem that has a sine in the denominator. You can move the function to the numerator by switching from the sine of the angle to the cosecant. Of course, nothing is ever too easy. Each of these reciprocal functions will be undefined where the original function is 0. In other words, because $\sin 0° = 0$, $\csc 0°$ would becomes $\frac{1}{0}$, which is undefined. In addition, because sine and cosine are always between -1 and 1, the values of secant and cosecant will always be either greater than or equal to 1 or less than or equal to -1.

Example 1: Let's find the cosecant, secant, and cotangent of the special angles. We just did them for 30°, now let's find them for 45°. We get:

$$\csc 45° = \frac{1}{\sin 45°} = \frac{1}{\sqrt{2}/2} = \sqrt{2}$$
$$\sec 45° = \frac{1}{\cos 45°} = \frac{1}{\sqrt{2}/2} = \sqrt{2}$$
$$\cot 45° = \frac{1}{\tan 45°} = \frac{1}{1} = 1.$$

Example 2: Now let's find the cosecant, secant, and cotangent of 60°. They are:

$$\csc 60° = \frac{1}{\sin 60°} = \frac{1}{\sqrt{3}/2} = \frac{2}{\sqrt{3}}$$

$$\sec 60° = \frac{1}{\cos 60°} = \frac{1}{1/2} = 2$$

$$\cot 60° = \frac{1}{\tan 60°} = \frac{1}{\sqrt{3}}.$$

Let's make a table for cosecant, secant, and cotangent, just as we did with the basic trig functions. Of course, if you just memorize the sine, cosine, and tangent of the special angles, you can always find the cosecant, secant, and cotangent by turning the values upside down.

	30°	45°	60°
Csc	2	$\sqrt{2}$	$\frac{2}{\sqrt{3}}$
Sec	$\frac{2}{\sqrt{3}}$	$\sqrt{2}$	2
Cot	$\sqrt{3}$	1	$\frac{1}{\sqrt{3}}$

Example 3: Now let's find the cosecant, secant, and cotangent of 0°.
Remember that $\sin 0° = 0$, so $\csc 0°$ is undefined.
Now let's repeat the process for secant: $\cos 0° = 1$, so $\sec 0° = 1$.
Finally, let's do cotangent: $\tan 0° = 0$, so $\cot 0°$ is undefined.

Example 4: Next, let's find the cosecant, secant, and cotangent of 90°.
We know that $\sin 90° = 1$, so $\csc 90° = 1$.
Next, $\cos 90° = 0$, so $\sec 90°$ is undefined.
Finding $\cot 90°$ is a little trickier. Remember how we found $\tan 90°$? It is $\frac{1}{0}$, which is undefined. But if $\tan 90° = \frac{1}{0}$, then $\cot 90° = \frac{0}{1} = 0$. So, even though the tangent is undefined, the cotangent has a value.

Example 5: Find the cosecant, secant, and cotangent of 180°.
First, $\sin 180° = 0$, so $\csc 180°$ is also undefined.
Next, $\cos 180° = -1$, so $\sec 180° = -1$.
And $\tan 180° = 0$, so $\cot 180°$ is undefined.

UNIT FIVE: The Reciprocal Functions

Example 6: Find the cosecant, secant, and cotangent of 270°.
First, $\sin 270° = -1$, so $\csc 270° = -1$.
Next, $\cos 270° = 0$, so $\sec 270°$ is undefined.
And finally, similar to finding how we found $\cot 90°$, we get $\cot 270° = 0$.
Are you ready to practice? Earlier, we found the sine, cosine, and tangent of a bunch of angles. Now we are going to find the cosecant, secant, and cotangent of the same angles.

Practice Problems

Practice problem 1: Find the cosecant, secant, and cotangent of 120°.
Practice problem 2: Find the cosecant, secant, and cotangent of 135°.
Practice problem 3: Find the cosecant, secant, and cotangent of 210°.
Practice problem 4: Find the cosecant, secant, and cotangent of 240°.
Practice problem 5: Find the cosecant, secant, and cotangent of 315°.
Practice problem 6: Find the cosecant, secant, and cotangent of 330°.
Practice problem 7: Find the cosecant, secant, and cotangent of 480°.
Practice problem 8: Find the cosecant, secant, and cotangent of 870°.
Practice problem 9: Find the cosecant, secant, and cotangent of −45°.
Practice problem 10: Find the cosecant, secant, and cotangent of −300°.

Solutions to the Practice Problems

Solution to practice problem 1: *Find the cosecant, secant, and cotangent of* 120°.
First, let's draw a picture so that we can find the reference angle:

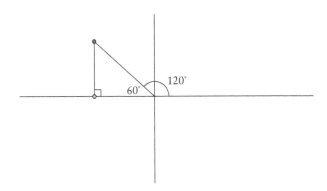

We can see that the acute angle made with the *x*-axis is 60° (the reference angle), so we need to find the cosecant, secant, and cotangent of 60°. Remember that because the sine will be positive in Quadrant II, the cosecant will be too. Also, because the cosine and tangent will be negative in Quadrant II, so too will the secant and cotangent be negative. We get:

$$\csc 120° = \frac{2}{\sqrt{3}}, \ \sec 120° = -2, \text{ and } \cot 120° = -\frac{1}{\sqrt{3}}.$$

Solution to practice problem 2: *Find the cosecant, secant, and cotangent of* 135°.

First, let's draw a picture so that we can find the reference angle:

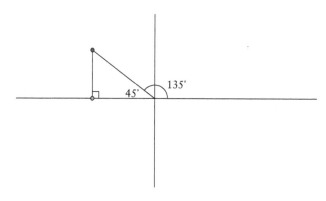

We can see that the acute angle made with the *x*-axis is 45° (the reference angle), so we need to find the cosecant, secant, and cotangent of 45°. Remember that because the sine will be positive in Quadrant II, the cosecant will be too. Also, because the cosine and tangent will be negative in Quadrant II, so too will the secant and cotangent be negative. We get:

$$\csc 135° = \sqrt{2}, \ \sec 135° = -\sqrt{2}, \text{ and } \cot 135° = -1.$$

Solution to practice problem 3: *Find the cosecant, secant, and cotangent of* 210°.

First, let's draw a picture so that we can find the reference angle:

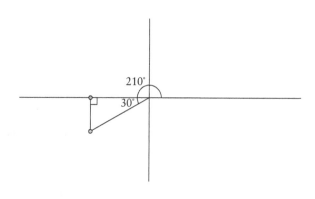

UNIT FIVE: The Reciprocal Functions   37

We can see that the acute angle with the *x*-axis is 30° (the reference angle), so we need to find the cosecant, secant, and cotangent of 30°. Remember that because the tangent will be positive in Quadrant III, the cotangent will be too. Also, because the sine and cosine will be negative in Quadrant III, so too will the cosecant and secant be negative. We get:

$$\csc 210° = -2, \ \sec 210° = -\frac{2}{\sqrt{3}}, \text{ and } \cot 210° = \sqrt{3}.$$

Solution to practice problem 4: *Find the cosecant, secant, and cotangent of* 240°.
First, let's draw a picture so that we can find the reference angle:

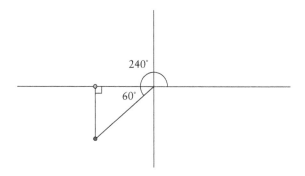

We can see that the acute angle made with the *x*-axis is 60° (the reference angle), so we need to find the cosecant, secant, and cotangent of 60°. Remember that because the tangent will be positive in Quadrant III, the cotangent will be too. Also, because the sine and cosine will be negative in Quadrant III, so too will the cosecant and secant be negative. We get:

$$\csc 240° = -\frac{2}{\sqrt{3}}, \ \sec 240° = -2, \text{ and } \cot 240° = \frac{1}{\sqrt{3}}.$$

Solution to practice problem 5: *Find the cosecant, secant, and cotangent of* 315°.
First, let's draw a picture so that we can find the reference angle:

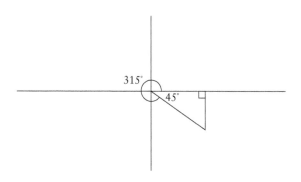

We can see that the acute angle made with the x-axis is 45° (the reference angle), so we need to find the cosecant, secant, and cotangent of 45°. Remember that because the cosine will be positive in Quadrant IV, the secant will be too. Also, because the sine and tangent will be negative in Quadrant IV, so too will the cosecant and cotangent be negative. We get:

$$\csc 315° = -\sqrt{2},\ \sec 315° = \sqrt{2},\ \text{and}\ \cot 315° = -1.$$

Solution to practice problem 6: *Find the cosecant, secant, and cotangent of 330°.*
First, let's draw a picture so that we can find the reference angle:

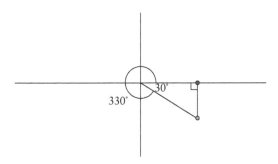

We can see that the acute angle with the x-axis is 30° (the reference angle), so we need to find the cosecant, secant, and cotangent of 30°. Remember that because the cosine will be positive in Quadrant IV, the secant will be too. Also, because the sine and tangent will be negative in Quadrant IV, so too will the cosecant and cotangent be negative. We get:

$$\csc 330° = -2,\ \sec 330° = \frac{2}{\sqrt{3}},\ \text{and}\ \cot 330° = -\sqrt{3}.$$

Solution to practice problem 7: *Find the cosecant, secant, and cotangent of 480°.*
First, let's draw a picture so that we can find the reference angle:

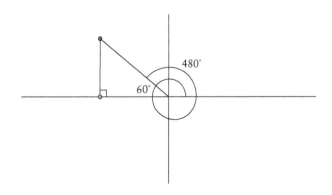

UNIT FIVE: The Reciprocal Functions

Here, we are going around the axes more than once (360°) and then continue another 120°, so we end up with the acute angle made with the x-axis as 60° (the reference angle). So we need to find the cosecant, secant, and cotangent of 60°. Remember that because the sine will be positive in Quadrant II, the cosecant will be too. Also, because the cosine and tangent will be negative in Quadrant II, so too will the secant and cotangent be negative. We get:

$$\csc 480° = \frac{2}{\sqrt{3}}, \sec 480° = -2, \text{ and } \cot 480° = -\frac{1}{\sqrt{3}}.$$

Solution to practice problem 8: *Find the cosecant, secant, and cotangent of* 870°.

First, let's draw a picture so that we can find the reference angle:

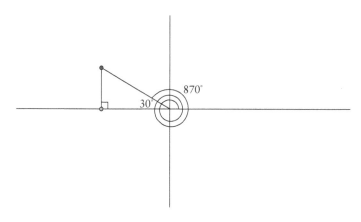

Here, we are going around the axes twice ($2 \cdot 360° = 720°$) and then continue another 150°. We can see that the acute angle made with the x-axis is 30° (the reference angle), so we need to find the cosecant, secant, and cotangent of 30°. Remember that because the sine will be positive in Quadrant II, the cosecant will be too. Also, because the cosine and tangent will be negative in Quadrant II, so too will the secant and cotangent be negative. We get:

$$\csc 870° = 2, \sec 870° = -\frac{2}{\sqrt{3}}, \text{ and } \cot 870° = -\sqrt{3}.$$

Solution to practice problem 9: *Find the cosecant, secant, and cotangent of* $-45°$.

Here, we find the reference angle by going around the axes in the opposite direction:

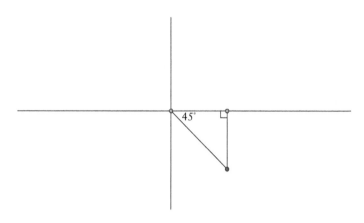

We can see that the acute angle made with the *x*-axis is $45°$ (the reference angle), so we need to find the cosecant, secant, and cotangent of $45°$. Remember that because the cosine will be positive in Quadrant IV, the secant will be too. Also, because the sine and tangent will be negative in Quadrant IV, so too will the cosecant and cotangent be negative. We get:

$$\csc(-45°) = -\sqrt{2}, \quad \sec(-45°) = \sqrt{2}, \text{ and } \cot(-45°) = -1.$$

Solution to practice problem 10: *Find the cosecant, secant, and cotangent of* $-300°$.

Here, we find the reference angle by going around the axes in the opposite direction, like so:

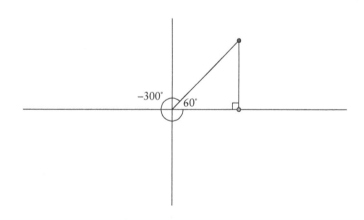

UNIT FIVE: The Reciprocal Functions 41

We can see that the acute angle made with the *x*-axis is 60° (the reference angle), so we need to find the cosecant, secant, and cotangent of 60°. Remember that the all of the functions will be positive because we are in Quadrant I. We get:

$$\csc(-300°) = \frac{2}{\sqrt{3}}, \ \sec(-300°) = 2, \text{ and } \cot(-300°) = \frac{1}{\sqrt{3}}.$$

UNIT SIX

Some Basic Trigonometry Problems

So far, we have learned all six trig ratios – sine, cosine, tangent, cotangent, secant, and cosecant – and we have learned how to find their values at a variety of angles. But, all of the angles that we have used so far have been the special angles. Now, we will learn how to find the trig ratios for any angle.

If you want to find the trig ratio of an angle, you simply use your calculator. For example, if you want to find $\sin 37°$, you plug $\sin 37°$ into your calculator. You should get approximately 0.6018. There are two very important things to keep in mind when using your calculator for trigonometry.

First, calculators will give you the choice of two modes – degrees or radians. (A few calculators also offer *grads*, but we will not work with grads in this book.) So, when you evaluate the trig ratio of an angle, **make sure that you are in the correct mode**! An easy way to check is to enter $\sin 30°$ on your calculator. If you are in *Degree Mode*, you will get 0.5. If you don't, then the calculator is in *Radian Mode*. Be careful! Many students mess up on exams by having their calculators in the wrong mode.

Second, when a calculator gives you the value of a trig function of an angle, it is giving you a decimal approximation. It is traditional to round to 4 decimal places, although you can feel free to use as many as you wish. Just keep in mind that your answer will be an approximation, not an exact value, unless you are finding the trig ratio of one of the special angles.

One type of problem that you will be asked to solve is, given an angle and a side of a triangle, find one of the other sides.

Example 1: In the right triangle ABC below, $\angle C$ is the right angle and $\angle A = 40°$. If $AB = 12$, how long is BC?

Figure 1

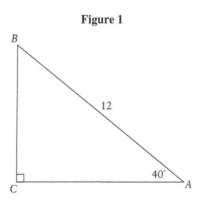

UNIT SIX: Some Basic Trigonometry Problems   43

Given AB, the hypotenuse of the triangle, we are looking for BC, which is the side opposite $\angle A$. This means that we can use sine to solve for BC because the sine of an angle is $\dfrac{opposite}{hypotenuse}$ (Remember SOHCAHTOA?). This gives us $\sin 40° = \dfrac{BC}{12}$.

We multiply across by 12 to get: $(12)(\sin 40°) = BC$.
Now, we can use our calculator to find $\sin 40° \approx 0.6428$.
Therefore, $(12)(0.6428) = BC$ and $BC \approx 7.71$.

Example 2: In the same triangle from Example 1, how long is AC?
Now we can use the cosine to find AC, because the cosine of an angle is $\dfrac{adjacent}{hypotenuse}$. This gives us $\cos 40° = \dfrac{AC}{12}$.

We multiply across by 12 to get: $(12)(\cos 40°) = AC$.
Now we can use our calculator to find $\cos 40° \approx 0.7660$.
Therefore, $(12)(0.7660) = AC$ and $AC \approx 9.19$.

We can check our answer with the Pythagorean Theorem. That is, does $AC^2 + BC^2 = AB^2$? $AC^2 + BC^2 = 7.71^2 + 9.19^2 \approx 143.9$. We know that $AB^2 = 12^2 = 144$. Allowing for rounding, our answers agree.

Example 3: In the right triangle ABC below, $\angle C$ is the right angle and $\angle A = 34°$. If $AC = 15$, how long is AB?

Figure 2

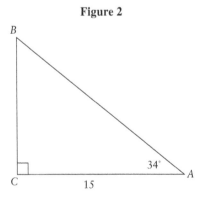

This time, we are given side AC, which is the side adjacent to $\angle A$, and we are looking for the hypotenuse of the triangle. This means that we can use cosine to solve for AC. This gives us $\cos 34° = \dfrac{15}{AB}$.

With a little algebra, we get: $AB = \dfrac{15}{\cos 34°}$.

Now we can use our calculator to find AB: $AB = \dfrac{15}{\cos 34} \approx 18.09$.

Let's do one with radians.

Example 4: In right triangle *DEF* below, ∠*F* is the right angle and $\angle D = \frac{\pi}{5}$. If *DF* = 6, find the length of side *EF*.

Figure 3

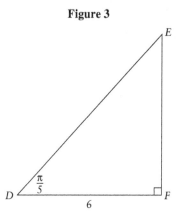

Here, we are looking for the side opposite ∠*D* and we are given the side adjacent to it. This means that we can use tangent to solve for *EF*. This gives us $\tan\frac{\pi}{5} = \frac{EF}{6}$. We multiply across by 6 to get: $6\tan\frac{\pi}{5} = EF$ (Notice that we don't really need parentheses). Now we use our calculator to find *EF*: $6\tan\frac{\pi}{5} = 6(.7265) \approx 4.36$.

Example 5: In the figure below, right triangle *ABC* has ∠*C* as the right angle and ∠*A* = 60°. If *BC* = 7, how long is *AB*?

Figure 4

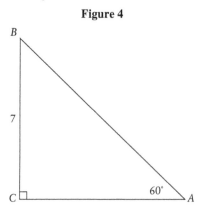

We are given side *BC*, the side opposite ∠*A*, and we are looking for the hypotenuse of the triangle. This means that we can use sine to solve for *AC*.

This gives us $\sin 60° = \frac{7}{AB}$.

UNIT SIX: Some Basic Trigonometry Problems

With a little algebra, we get: $AB = \dfrac{7}{\sin 60°}$.

This time, we don't have to use our calculator because we know the *exact* value of AB. We learned in the unit on Special Triangles that $\sin 60° = \dfrac{\sqrt{3}}{2}$.

$$AB = \dfrac{7}{\sqrt{3}/2} = \dfrac{14}{\sqrt{3}}$$

Another type of problem that you will be asked to solve is to find a trig ratio given one of the other trig ratios for a triangles.

Example 6: Given a right triangle ABC with $\angle C$ as the right angle and $\sin A = \dfrac{4}{5}$, find $\cos A$.

Let's draw a picture.

Figure 5

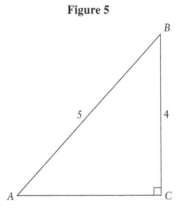

Notice that we labeled side $BC = 4$ and the hypotenuse $AB = 5$. We can do this because we know that, BC is the side opposite $\angle A$, and AB is the hypotenuse, and $\sin A = \dfrac{opposite}{hypotenuse}$. Of course, we could have used any two numbers for the sides whose ratio is $\dfrac{4}{5}$, but why make life hard for ourselves?

Now we can use the Pythagorean Theorem to solve for the missing side:

$$AC^2 + BC^2 = AB^2$$

$$AC^2 + 4^2 = 5^2$$

$$AC = 3.$$

Now we can use the fact that $\cos A = \dfrac{adjacent}{hypotenuse}$ to find $\cos A = \dfrac{3}{5}$.

Let's do another one.

Example 7: Given right triangle PRS with $\angle S$ as the right angle, and $\tan R = \dfrac{3}{2}$, find $\cos R$.

Figure 6

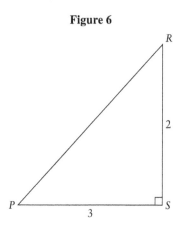

Notice that we labeled side $PS = 3$ and side $RS = 2$. We can do this because we know that PS is the side opposite $\angle R$, RS is the side adjacent to $\angle R$, and $\tan R = \dfrac{opposite}{adjacent}$. As in the last example, we could have used any two numbers for the sides whose ratio is $\dfrac{3}{2}$ but it is generally best to choose the simplest two numbers.

Now we can use the Pythagorean Theorem to solve for the missing hypotenuse:

$$PS^2 + RS^2 = PR^2$$
$$3^2 + 2^2 = PR^2$$
$$PR = \sqrt{13}.$$

Notice that we don't get an integer this time but that doesn't matter. Now we can use the fact that $\cos R = \dfrac{adjacent}{hypotenuse}$ to find $\cos R = \dfrac{2}{\sqrt{13}}$.

Now let's make things a little more complicated!

Example 8: If $\sin x = \dfrac{5}{13}$ and x is in Quadrant II, find $\tan x$.

Let's draw a picture:

UNIT SIX: Some Basic Trigonometry Problems   47

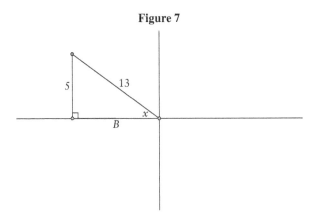

Figure 7

Remember that when we are in Quadrant II, the sine of an angle is positive, but the tangent and cosine are negative (**A**ll **S**tudents **T**ry **C**andy).

Here is how we solve this. First, as in the previous two examples, we use the Pythagorean Theorem to solve for the missing side B:

$$5^2 + B^2 = 13^2$$
$$B = \pm 12.$$

The answer is actually $B = -12$ because B is on the negative side of the x-axis. This is why the various trig functions are positive or negative in the different quadrants, as we learned previously. If you want, you can ignore the minus sign when using the Pythagorean Theorem, but *don't forget* to use the correct sign in your final answer!

Now, we can find $\tan x = -\dfrac{5}{12}$.

Let's do another one.

Example 9: If $\tan \theta = \dfrac{3}{7}$ and θ is in Quadrant III, find $\sin \theta$ and $\sec \theta$.

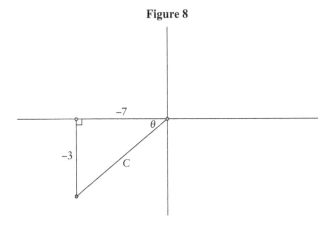

Figure 8

Note that 3 and 7 are negative because they are located on the negative x and y-axes. As before, you could ignore the signs and adjust the answer at the end. Remember that when we are in Quadrant III, the tangent of an angle is positive, but the sine and cosine are negative (**A**ll **S**tudents **T**ry **C**andy).

We can use the Pythagorean Theorem to solve for the hypotenuse C:

$$3^2 + 7^2 = C^2$$

$C = \sqrt{58}$. (Hypotenuses are always treated as positive.)

Now we can find $\sin\theta = -\dfrac{3}{\sqrt{58}}$ and $\sec\theta = -\dfrac{\sqrt{58}}{7}$.

Example 10: If $\sec\theta = \dfrac{9}{5}$ and $\sin\theta < 0$, find $\cot\theta$.

How do we figure out which quadrant to place the triangle? It's simple. We are told that $\sec\theta$ is positive and $\sin\theta$ is negative. We know that cosine is positive in Quadrants I and IV, which means that secant is too. We also know that sine is negative in Quadrants III and IV. The quadrant that satisfies both of these conditions is Quadrant IV.

Let's draw a picture.

Figure 9

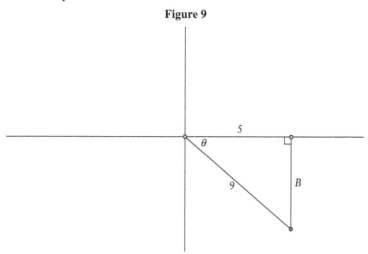

We can use the Pythagorean Theorem to solve for the opposite side B:

$$5^2 + B^2 = 9^2$$

$B = \sqrt{56}$. (We are ignoring the minus sign because we know that the tangent must be negative in quadrant IV.)

Now we can find $\tan\theta = -\dfrac{\sqrt{56}}{5}$ and $\cot\theta = -\dfrac{5}{\sqrt{56}}$.

Now let's look at some of the typical word problems that you will be expected to solve using Trigonometry.

UNIT SIX: Some Basic Trigonometry Problems   49

Example 11: Suppose you are 50 feet (ft) from the base of a tree that is standing straight up. You measure the angle of elevation to the top of the tree as 52°. How tall is the tree to the nearest foot?

The tree makes a right angle with the ground, so we can construct a right triangle with the information that we are given. Let's draw a picture.

Figure 10

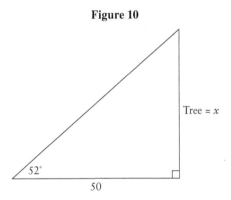

We know the distance to the base of the tree, which is the side of the triangle adjacent to the angle of 52°. We are looking for the height of the tree, which is the side opposite the angle. This means that we can use the tangent of the angle to find the height of the tree because tangent is $\frac{opposite}{adjacent}$ (remember SOHCAHTOA?).

Call the height of the tree x, and we can make the equation: $\tan 52° = \frac{x}{50}$.

Next, multiply across by 50 to get: $(50)(\tan 52°) = x$.

Using the calculator, $\tan 52° \approx 1.2799$, gives us: $(50)(1.2799) = x$ and $x = 63.995 \approx 64$ ft.

We solved the problem by using the given information to construct a right triangle. Then we used trigonometry to solve the problem. Let's do another.

Example 12: A road makes an angle of 7° with the flat ground and eventually rises to a height of 400 ft above its initial height. How long is the road to the nearest foot?

Once again, we can draw a right triangle and put in the information that we have been given.

Figure 11

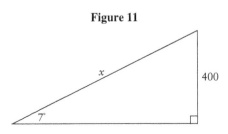

Here, we are given the side opposite the 7°-angle, and we are looking for the length of the road, which is the hypotenuse x. This means that we can use the sine of the angle to find x.

We get: $\sin 7° = \dfrac{400}{x}$.

Now, we can rearrange to solve for x: $x = \dfrac{400}{\sin 7°}$.

Using the calculator, $\sin 7° \approx 0.1219$, giving us: $x = \dfrac{400}{0.1219} \approx 3281$ ft.

Example 13: A boat is floating near a dock that is higher than the water and the front end of the boat is attached to the end of the dock by a rope that is 22 ft long. If the angle that the rope makes with the dock is 72°, how high above the water is the dock, and how far away is the front of the boat from the dock?

Let's draw a picture. We call the height of the dock y, and the distance that the boat is from the dock, x.

Figure 12

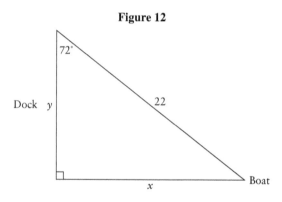

First, let's find y, the height of the dock. We can use cosine because we have the hypotenuse of a right triangle (the rope), and we are looking for the side adjacent to the 72°-angle.

We get: $\cos 72° = \dfrac{y}{22}$.

Multiply both sides by 22: $22 \cos 72° = y$.
We get that the dock is approximately 6.80 ft above the water.
Now let's find x. This time we will use sine to solve for x.

We get: $\sin 72° = \dfrac{x}{22}$.

Multiply both sides by 22: $22 \sin 72° = x$.
We get that the boat is approximately 20.92 ft from the dock.

By the way, once we had found side y, we could have used the Pythagorean Theorem to solve for x: $22^2 - 6.8^2 = x^2$ and $x \approx 20.92$.

UNIT SIX: Some Basic Trigonometry Problems 51

Did you notice that we could often use the Pythagorean Theorem, instead of a trig function, to solve a problem? This is because there is a relationship between Trigonometry and the Pythagorean Theorem. We will be exploring this later in this book.

Here is one last type of problem. This will require us to use two different trig ratios to solve the problem.

Example 14: From the top of a 60-ft tall vertical observation tower, a forest ranger observes a deer moving toward her. If the angle of depression from the ranger to the deer is initially 20° and a minute later is 35°, how far did the deer move during that minute?

Let's draw a picture. D is the initial location of the deer and C is its location one minute later.

Figure 13

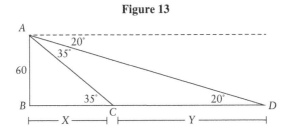

Notice that $\angle ADB = 20°$ and $\angle ACB = 35°$ because the alternate interior angles formed by a transversal of parallel lines are congruent. Now, we can use tangents to solve the problem. First, we find from $\triangle ABC$ that $\tan 35° = \dfrac{60}{x}$, which we can rearrange to $x = \dfrac{60}{\tan 35°}$. Next, from $\triangle ADB$ we can find that $\tan 20° = \dfrac{60}{x+y}$, which we can rearrange to $x + y = \dfrac{60}{\tan 20°}$.

Now for some algebra!
Substitute the expression for x in the first equation into the second one:

$$\dfrac{60}{\tan 35°} + y = \dfrac{60}{\tan 20°}.$$

Isolate y: $y = \dfrac{60}{\tan 20°} - \dfrac{60}{\tan 35°}$.

Now we use our calculator: $y \approx 164.8486 - 85.6889 \approx 79.2$ ft.

Are you ready for some practice problems?

Practice Problems

Practice problem 1: In right triangle *ABC* below, ∠*C* is the right angle, and ∠*A* = 23°. If *AB* = 19, how long is *BC*?

Figure 14

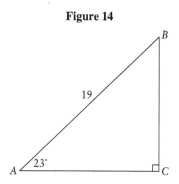

Practice problem 2: In the right triangle *ABC* below, ∠*C* is the right angle and ∠*B* = 77°. If *AC* = 11, how long is *AB*?

Figure 15

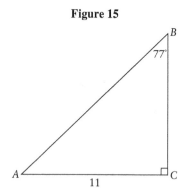

Practice problem 3: In the right triangle *PRS* below, ∠*S* is the right angle and ∠*P* = 19°. If *PS* = 20, how long is *RS*?

UNIT SIX: Some Basic Trigonometry Problems

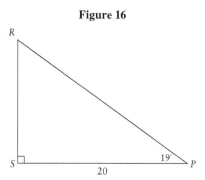

Figure 16

Practice problem 4: In the right triangle *DEF* below, ∠*F* is the right angle and ∠*D* = 38°. If *DE* = 100, how long is *DF*?

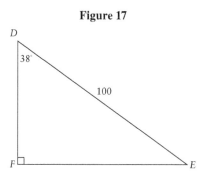

Figure 17

Practice problem 5: In the right triangle *XYZ* below, ∠*Z* is the right angle and ∠*Y* = 40°. If *XZ* = 10, how long is *YZ*?

Figure 18

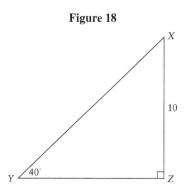

Practice problem 6: Given right triangle ABC with $\angle C$ as the right angle and $\sin A = \dfrac{7}{25}$, find $\cos A$.

Practice problem 7: Given right triangle ABC with $\angle C$ as the right angle and $\cos A = \dfrac{8}{17}$, find $\cot A$.

Practice problem 8: Given right triangle LMN with $\angle N$ as the right angle and $\sec M = \dfrac{9}{4}$, find $\sin M$.

Practice problem 9: If $\sin x = \dfrac{9}{41}$ and x is in Quadrant I, find $\cot x$.

Practice problem 10: If $\csc x = \dfrac{13}{12}$ and x is in Quadrant II, find $\cos x$.

Practice problem 11: If $\sin x = -\dfrac{9}{10}$ and x is in Quadrant IV, find $\sec x$.

Practice problem 12: If $\cos x = -\dfrac{4}{11}$ and x is in Quadrant III, find $\tan x$.

Practice problem 13: If $\sin \theta = \dfrac{4}{13}$ and $\tan \theta < 0$, find $\sec \theta$.

Practice problem 14: If $\sec \theta = \dfrac{14}{5}$ and $\cot \theta < 0$, find $\sin \theta$.

Practice problem 15: If $\csc \theta = -\dfrac{7}{3}$ and $\tan \theta > 0$, find the other five trig ratios of θ.

Practice problem 16: If $\cos \theta = -\dfrac{2}{9}$ and $\sin \theta > 0$, find the other five trig ratios of θ.

Practice problem 17: The angle of elevation to the top of a vertical tower is 55° from a point 200 ft away from the base of the tower on level ground. What is the height of the tower?

Practice problem 18: A support wire connects the top of a vertical tent pole to a point on level ground 16 ft from the base of the pole. If the wire makes an angle of 36° with the ground, find the length of the wire.

Practice problem 19: A ship is located off shore at point P, and R is the closest point on the straight shoreline, where $PR \perp RQ$. Point Q is located 3.2 miles (mi) down the shoreline from R. If $\angle P = 44°$, how far is the ship from R?

Practice problem 20: A person stands on the far side of a street and looks at a building set back from the other side of the street. He measures the angle of elevation to the top of the building as 54°. He crosses the street, which is 30 ft wide, and now measures the angle of elevation as 71°. Find the height of the building.

UNIT SIX: Some Basic Trigonometry Problems

Practice problem 21: Two people stand 100 ft apart and look up at a balloon in the sky. Person A measures the angle of elevation to the balloon as 66°. Person B measures the angle of elevation to the balloon as 76°. How high is the balloon?

Solutions to the Practice Problems

Solution to practice problem 1: *In the right triangle ABC below, $\angle C$ is the right angle and $\angle A = 23°$. If $AB = 19$, how long is BC?*

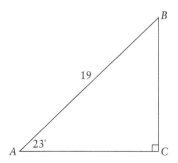

AB is the hypotenuse of the triangle and we are looking for BC, which is the side opposite $\angle A$. This means that we can use sine to solve for BC. This gives us $\sin 23° = \dfrac{BC}{19}$.

We multiply across by 19 to get: $19 \sin 23° = BC$.
Now we use our calculator to find $\sin 23° \approx 0.3907$.
Therefore, $19(0.3907) = BC$ and $BC \approx 7.42$.

Solution to practice problem 2: *In the right triangle ABC below, $\angle C$ is the right angle and $\angle B = 77°$. If $AC = 11$, how long is AB?*

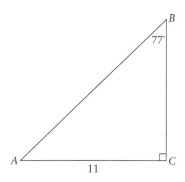

Here, we have AC, which is the side opposite $\angle A$, and we are looking for AB, the hypotenuse of the triangle. This means that we can use sine to solve for AB. We get: $\sin 77° = \dfrac{11}{AB}$.

We can rearrange the equation to get: $AB = \dfrac{11}{\sin 77°}$.

Now we use our calculator to find: $AB = \dfrac{11}{\sin 77°} \approx 11.3$.

Solution to practice problem 3: *In the right triangle PRS below, $\angle S$ is the right angle and $\angle P = 19°$. If $PS = 20$, how long is RS?*

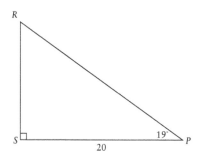

Here, we have PS, which is the side adjacent to $\angle P$, and we are looking for RS, which is the side opposite $\angle P$. This means that we can use tangent to solve for RS.
We get: $\tan 19° = \dfrac{RS}{20}$.
Multiply across by 20 to get: $20 \tan 19° = RS$.
Now we use our calculator to find: $RS = 20 \tan 19° \approx 6.89$.

Solution to practice problem 4: *In the right triangle DEF below, $\angle F$ is the right angle and $\angle D = 38°$. If $DE = 100$, how long is DF?*

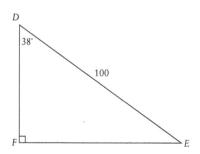

UNIT SIX: Some Basic Trigonometry Problems 57

Here, we have *DE*, which is the hypotenuse, and we are looking for *DF*, which is the side adjacent to ∠*D*. This means that we can use cosine to solve for *DF*.

We get: $\cos 38° = \dfrac{DF}{100}$.

Multiply across by 100 to get: $100 \cos 38° = DF$.
Now we use our calculator to find: $DF = 100 \cos 38° \approx 78.8$.

Solution to practice problem 5: *In the right triangle XYZ below, ∠Z is the right angle and ∠Y = 40°. If XZ = 10, how long is YZ?*

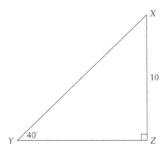

Here, we have *XZ*, which is the side opposite ∠*Y*, and we are looking for *YZ*, which is the side adjacent to ∠*Y*. This means that we can use tangent to solve for *YZ*.

We get: $\tan 40° = \dfrac{10}{YZ}$.

Rearrange to get: $YZ = \dfrac{10}{\tan 40°}$.

Now we use our calculator to find: $YZ = \dfrac{10}{\tan 40°} \approx 11.92$.

Solution to practice problem 6: *Given a right triangle ABC with ∠C as the right angle and* $\sin A = \dfrac{7}{25}$, *find* $\cos A$.

Let's draw a picture.

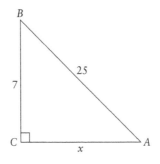

Notice that we labeled side $BC = 7$ and the hypotenuse $AB = 25$. We can do this because we know that $\sin A = \dfrac{opposite}{hypotenuse}$, BC is the side opposite $\angle A$, and AB is the hypotenuse. We could have used any two numbers for the sides whose ratio is $\dfrac{7}{25}$ and we chose the easiest pair of numbers. We labeled the unknown side x.

Now we can use the Pythagorean Theorem to solve for x:

$$7^2 + x^2 = 25^2$$
$$x = 24$$

Now we can use the fact that $\cos A = \dfrac{adjacent}{hypotenuse}$ to find $\cos A = \dfrac{24}{25}$.

Solution to practice problem 7: *Given right triangle ABC with $\angle C$ as the right angle and $\cos A = \dfrac{8}{17}$, find $\cot A$.*

Let's draw a picture.

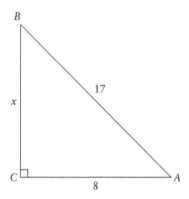

We labeled side $AC = 8$ and the hypotenuse $AB = 17$. We can do this because $\cos A = \dfrac{adjacent}{hypotenuse}$. We labeled the unknown side x.

Now we can use the Pythagorean Theorem to solve for x:

$$8^2 + x^2 = 17^2$$
$$x = 15$$

Now we can use the fact that $\cot A = \dfrac{adjacent}{opposite}$ to find $\cot A = \dfrac{8}{15}$.

UNIT SIX: Some Basic Trigonometry Problems

Solution to practice problem 8: *Given a right triangle LMN with ∠N as the right angle, and* $\sec M = \dfrac{9}{4}$, *find* $\sin M$.

Let's draw a picture.

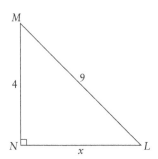

We labeled side $MN = 4$ and the hypotenuse $LM = 9$. We can do this because $\sec M = \dfrac{hypotenuse}{adjacent}$. We labeled the unknown side x.

Now we can use the Pythagorean Theorem to solve for x:

$$4^2 + x^2 = 9^2$$
$$x = \sqrt{65}$$

Now we can use the fact that $\sin M = \dfrac{opposite}{hypotenuse}$ to find $\sin M = \dfrac{\sqrt{65}}{9}$.

Solution to practice problem 9: *If* $\sin x = \dfrac{9}{41}$ *and x is in Quadrant I, find* $\cot x$.
Let's draw a picture:

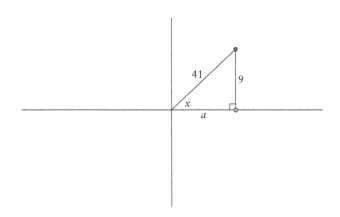

Remember that when we are in Quadrant I, all of the trig functions are positive (**A**ll **S**tudents **T**ry **C**andy). We can use the Pythagorean Theorem to solve for the missing side, which we have labeled a:

$$9^2 + a^2 = 41^2$$

$$a = \pm 40.$$

The answer is actually $a = 40$ because side a is on the x-axis.

We then find $\cot x = \dfrac{40}{9}$.

Solution to practice problem 10: *If* $\csc x = \dfrac{13}{12}$ *and x is in Quadrant II, find* $\cos x$. Let's draw a picture.

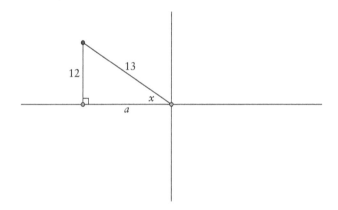

Remember that when we are in Quadrant II, the sine of an angle is positive, but the tangent and cosine are negative (**A**ll **S**tudents **T**ry **C**andy).

We use the Pythagorean Theorem to solve for the missing side a:

$$a^2 + 12^2 = 13^2$$
$$a = \pm 5.$$

The answer is actually $a = -5$ because side a is on the negative x-axis.

Now we can find $\cos x = -\dfrac{5}{13}$.

Solution to practice problem 11: *If* $\sin x = -\dfrac{9}{10}$ *and x is in Quadrant IV, find* $\sec x$.

Let's draw a picture.

UNIT SIX: Some Basic Trigonometry Problems   61

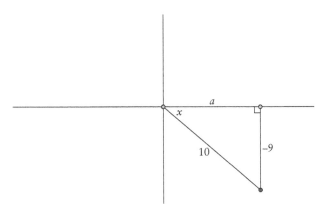

Remember that when we are in Quadrant IV, the cosine of an angle is positive, but the tangent and sine are negative (**A**ll **S**tudents **T**ry **C**andy).

We use the Pythagorean Theorem to solve for the missing side a:

$$a^2 + 9^2 = 10^2$$
$$a = \pm\sqrt{19}.$$

The answer is actually $a = \sqrt{19}$ because side a is on the positive x-axis.

Now we can find $\sec x = \dfrac{10}{\sqrt{19}}$.

Solution to practice problem 12: *If* $\cos x = -\dfrac{4}{11}$ *and x is in Quadrant III, find* $\tan x$.

Let's draw a picture.

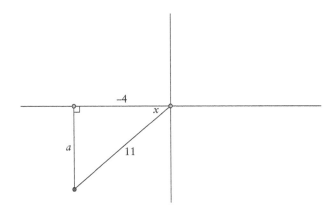

Remember that when we are in Quadrant III, the tangent of an angle is positive, but the sine and cosine are negative (**A**ll **S**tudents **T**ry **C**andy).

We use the Pythagorean Theorem to solve for the missing side a:

$$a^2 + 4^2 = 11^2$$
$$a = \pm\sqrt{105}.$$

The answer is actually $a = -\sqrt{105}$ because a is on the negative y-axis.

Now we can find $\tan x = \dfrac{\sqrt{105}}{4}$.

Solution to practice problem 13: *If* $\sin\theta = \dfrac{4}{13}$ *and* $\tan\theta < 0$, *find* $\sec\theta$.

We are told that $\sin\theta$ is positive and $\tan\theta$ is negative. We know that sine is positive in Quadrants I and II. We also know that tangent is negative in Quadrants II and IV. The quadrant that satisfies both of these conditions is Quadrant II.

Let's draw a picture.

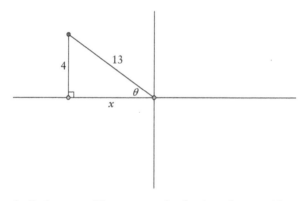

We can use the Pythagorean Theorem to solve for the unknown side x:

$$x^2 + 4^2 = 13^2$$
$$x = \sqrt{153}.$$

We are ignoring the minus sign because we know that the secant must be negative. (We are in Quadrant II, where cosine and secant are negative.)

Now we can find $\sec\theta = -\dfrac{13}{\sqrt{153}}$.

Solution to practice problem 14: *If* $\sec\theta = \dfrac{14}{5}$ *and* $\cot\theta < 0$, *find* $\sin\theta$.

We are told that $\sec\theta$ is positive and $\cot\theta$ is negative. We know that cosine (and hence, secant) is positive in Quadrants I and IV. We also know that tangent

UNIT SIX: Some Basic Trigonometry Problems   63

(and hence, cotangent) is negative in Quadrants II and IV. The quadrant that satisfies both of these conditions is Quadrant IV.
Let's draw a picture.

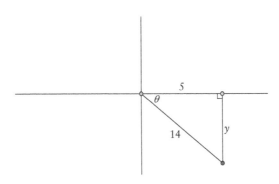

We can use the Pythagorean Theorem to solve for the unknown side y:

$$5^2 + y^2 = 14^2$$

$$y = \sqrt{171}.$$

We are ignoring the minus sign because we know that the sine must be negative. (We are in quadrant IV, where sine is negative.)

Now we can find $\sin\theta = -\dfrac{\sqrt{171}}{14}$.

Solution to practice problem 15: *If* $\csc\theta = -\dfrac{7}{3}$ *and* $\tan\theta > 0$, *find the other five trig ratios of* θ.

We are told that $\csc\theta$ is negative and $\tan\theta$ is positive. We know that sine (and hence, cosecant) is negative in Quadrants III and IV. We also know that tangent is positive in Quadrants I and III. The quadrant that satisfies both of these conditions is Quadrant III.
Let's draw a picture.

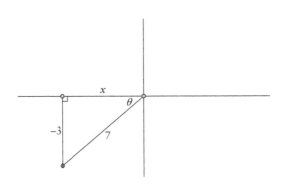

We can use the Pythagorean Theorem to solve for the unknown side x:

$$x^2 + 3^2 = 7^2$$

$$x = \sqrt{40}.$$

Now we can find the other five ratios. Remember that the cotangent will be positive and that the other four ratios will be negative. We get: $\sin\theta = -\frac{3}{7}$. (By the way, we could have found this by turning the cosecant upside down.)

$$\cos\theta = -\frac{\sqrt{40}}{7}$$

$$\sec\theta = -\frac{7}{\sqrt{40}}$$

$$\tan\theta = \frac{3}{\sqrt{40}}$$

$$\cot\theta = \frac{\sqrt{40}}{3}$$

Solution to practice problem 16: If $\cos\theta = -\frac{2}{9}$ and $\sin\theta > 0$, find the other five trig ratios of θ.

We are told that $\cos\theta$ is negative and $\sin\theta$ is positive. We know that cosine is negative in quadrants II and III. We also know that sine is positive in Quadrants I and II. The quadrant that satisfies both of these conditions is Quadrant II.

Let's draw a picture.

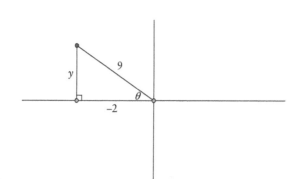

UNIT SIX: Some Basic Trigonometry Problems

We can use the Pythagorean Theorem to solve for the unknown side y:

$$2^2 + y^2 = 9^2$$

$y = \sqrt{77}$. Now we can find the other five ratios. Remember that the cosecant will be positive and that the other four ratios will be negative. We get:

$$\sin\theta = \frac{\sqrt{77}}{9}$$

$$\csc\theta = \frac{9}{\sqrt{77}}$$

$$\sec\theta = -\frac{9}{2} \text{ (By the way, we could have found this by turning the cosecant upside down.)}$$

$$\tan\theta = -\frac{\sqrt{77}}{2}$$

$$\cot\theta = -\frac{2}{\sqrt{77}}.$$

Solution to practice problem 17: *The angle of elevation to the top of a vertical tower is 55° from a point 200 ft away from the base of the tower on level ground. What is the height of the tower?*

The tower makes a right angle with the ground, so we can construct a right triangle with the information that we are given.

Let's draw a picture.

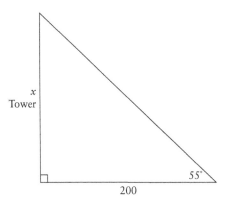

We know the distance to the base of the tower, which is the side of the triangle adjacent to the angle of $55°$. We are looking for the height of the tower, which is the side opposite the angle. This means that we can use the tangent of $55°$ to find the height of the tower. Call the height of the tower x, and we can make the equation: $\tan 55° = \dfrac{x}{200}$.

Next, multiply across by 200 to get: $200 \tan 55° = x$.
Using the calculator, we get: $200 \tan 55° = 285.6$ ft.

Solution to practice problem 18: *A support wire connects the top of a vertical tent pole to a point on level ground 16 ft from the base of the pole. If the wire makes an angle of $36°$ with the ground, find the length of the wire.*

We can draw a right triangle and put in the information that we have been given.

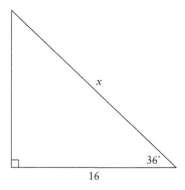

Here, we are given the side adjacent to the $36°$-angle, and we are looking for the length of the wire, which is the hypotenuse x. This means that we can use the cosine of $36°$ to find x.

We get: $\cos 36° = \dfrac{16}{x}$.

Now, we can rearrange to solve for x: $x = \dfrac{16}{\cos 36°}$.

Using the calculator, we get $\dfrac{16}{\cos 36°} = x = 19.78$ ft.

Solution to practice problem 19: *A ship is located off shore at point P, and R is the closest point on the straight shoreline, where $PR \perp RQ$. Point Q is located 3.2 miles (mi) down the shoreline from R. If $\angle P = 44°$, how far is the ship from R?*

We can draw a right triangle and put in the information that we have been given.

UNIT SIX: Some Basic Trigonometry Problems   67

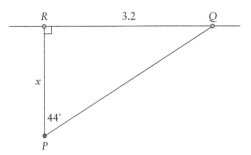

Here, we are given the side opposite ∠P, and we are looking for the distance x, which is the side adjacent to ∠P. This means that we can use the tangent of 44° to find x:

$$\tan 44° = \frac{3.2}{x}.$$

Now, we can rearrange to solve for x: $x = \frac{3.2}{\tan 44°}$.

Using the calculator, we get $x = \frac{3.2}{\tan 44°} \approx 3.3$ mi.

(By the way, since the acute angles in an isosceles right triangle are both 45°, and here, the angles are 44° and 46°, we were expecting the distance to be very close to 3.2 mi.)

Solution to practice problem 20: *A person stands on the far side of a street and looks at a building set back from the other side of the street. He measures the angle of elevation to the top of the building as 54°. He crosses the street, which is 30 ft wide, and now measures the angle of elevation as 71°. Find the height of the building.*

We can construct a pair of right triangles to solve the problem. Let's draw a picture.

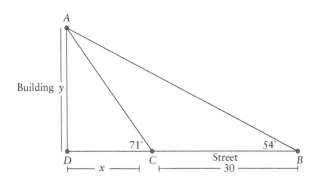

We are looking for y, the height of the building. and we can use tangents to solve the problem. First, we find from $\triangle ABD$ that $\tan 54° = \dfrac{y}{x+30}$, which we can rearrange to be $\tan 54°(x+30) = y$. Next, from $\triangle ACD$ we can find that $\tan 71° = \dfrac{y}{x}$, which we can rearrange to be $x\tan 71° = y$.

Now for some algebra!
Substitute the expression for y in the first equation into the second one: $\tan 54°(x+30) = x\tan 71°$.
  Distribute: $x\tan 54° + 30\tan 54° = x\tan 71°$.
  Group the x terms: $30\tan 54° = x\tan 71° - x\tan 54°$.
  Factor out x: $30\tan 54° = x(\tan 71° - \tan 54°)$.
  Now we can use our calculator: $41.291 = 1.528x$.

$$x \approx \dfrac{41.291}{1.528} \approx 27.02.$$

We found x but remember that we are looking for y!
We get: $y = x\tan 71° = 27.02\tan 71° \approx 78.5$ ft.

Solution to practice problem 21: *Two people stand 100 ft apart and look up at a balloon in the sky. Person A measures the angle of elevation to the balloon as 66°. Person B measures the angle of elevation to the balloon as 76°. How high is the balloon?*

We can construct a pair of right triangles to solve the problem. Let's draw a picture.

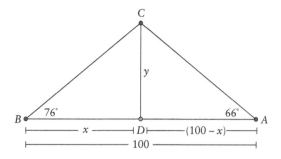

We are looking for y, the height of the balloon. and again, we can use tangents to solve the problem. First, we find from $\triangle BCD$ that $\tan 76° = \dfrac{y}{x}$, which we can rearrange to $x\tan 76° = y$. Next, from $\triangle ACD$ we can find that $\tan 66° = \dfrac{y}{100-x}$, which we can rearrange to $\tan 66°(100-x) = y$.

UNIT SIX: Some Basic Trigonometry Problems 69

Now for some algebra!
Substitute the expression for y in the first equation into the second one: $\tan 66°\,(100-x) = x\tan 76°$.
Distribute: $100\tan 66° - x\tan 66° = x\tan 76°$.
Group the x terms: $100\tan 66° = x\tan 76° + x\tan 66°$.
Factor out x: $100\tan 66° = x(\tan 76° + \tan 66°)$.
Now we can use our calculator: $224.604 = 6.257x$.

$$x \approx \frac{224.604}{6.257} \approx 35.896$$

Now to find y: $y = x\tan 76° = 35.896\tan 76° \approx 143.97$ ft.

UNIT SEVEN

Sine and Cosine Graphs

Now we are going to learn how to graph the sine function. By the way, throughout this unit we will use radians rather than degrees. We do this because it makes the scaling of the graphs easier. That is, the values on the x-axis will range from 0 to 2π (which is approximately 6.28), rather than from 0° to 360°. First, let's make a table of the value of $\sin x$ as x goes from 0 to 2π radians.

x	0	$\dfrac{\pi}{6}$	$\dfrac{\pi}{4}$	$\dfrac{\pi}{3}$	$\dfrac{\pi}{2}$	$\dfrac{2\pi}{3}$	$\dfrac{3\pi}{4}$	$\dfrac{5\pi}{6}$	π	$\dfrac{7\pi}{6}$	$\dfrac{5\pi}{4}$	$\dfrac{4\pi}{3}$	$\dfrac{3\pi}{2}$	$\dfrac{5\pi}{3}$	$\dfrac{7\pi}{4}$	$\dfrac{11\pi}{6}$	2π
$\sin x$	0	$\dfrac{1}{2}$	$\dfrac{\sqrt{2}}{2}$	$\dfrac{\sqrt{3}}{2}$	1	$\dfrac{\sqrt{3}}{2}$	$\dfrac{\sqrt{2}}{2}$	$\dfrac{1}{2}$	0	$-\dfrac{1}{2}$	$-\dfrac{\sqrt{2}}{2}$	$-\dfrac{\sqrt{3}}{2}$	-1	$-\dfrac{\sqrt{3}}{2}$	$-\dfrac{\sqrt{2}}{2}$	$-\dfrac{1}{2}$	0

Notice how the values of sine start at 0, rise to 1, go back to 0, then go down to −1 and then rise back to 0. Remember that we find these values of sine by going around the unit circle. It should make sense that they rise and fall (imagine yourself on a Ferris Wheel), and that they repeat.

Now let's graph the values of sine. The x-axis will be for the values of the angles and the y-axis will be for the values of $\sin x$.

Figure 1

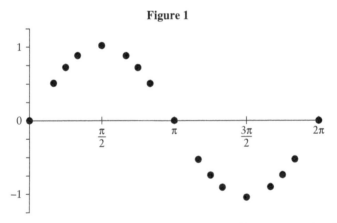

Remember that sine can have any value from −1 to 1, so if we were to plot all of the possible values, we would get a smooth curve:

UNIT SEVEN: Sine and Cosine Graphs

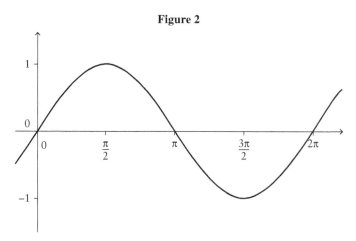

Figure 2

See how it is shaped like a wave? In fact, sine graphs are often referred to as *sine waves* for this reason. Remember that we can go around the unit circle as many times as we wish, both in the positive and negative directions, so the sine graph will repeat, creating an infinitely long graph. It is traditional to graph only one full wavelength of the graph and, absent any other information, to graph the curve beginning at the origin and extending to the right 2π units. The length of one full curve is referred to as the *period* of the graph. In other words, one sine wave covers an interval of length 2π before it repeats itself. This also means that $\sin(x + 2\pi) = \sin x$.

Here are the important things to remember about a sine graph:

- Its minimum y-value is -1 and its maximum y-value is 1.
- The period of one wavelength is 2π.
- We refer to the *beginning* of the graph as the origin.
- The midline of the graph is the x-axis.

These will be important when we graph more complex sine and cosine curves, so make sure that you understand them.

Now let's draw a cosine curve. Remember that cosine and sine take on the same y values, but for different x values. Let's make a table for cosine:

x	0	$\dfrac{\pi}{6}$	$\dfrac{\pi}{4}$	$\dfrac{\pi}{3}$	$\dfrac{\pi}{2}$	$\dfrac{2\pi}{3}$	$\dfrac{3\pi}{4}$	$\dfrac{5\pi}{6}$	π	$\dfrac{7\pi}{6}$	$\dfrac{5\pi}{4}$	$\dfrac{4\pi}{3}$	$\dfrac{3\pi}{2}$	$\dfrac{5\pi}{3}$	$\dfrac{7\pi}{4}$	$\dfrac{11\pi}{6}$	2π
$\cos x$	1	$\dfrac{\sqrt{3}}{2}$	$\dfrac{\sqrt{2}}{2}$	$\dfrac{1}{2}$	0	$-\dfrac{1}{2}$	$-\dfrac{\sqrt{2}}{2}$	$-\dfrac{\sqrt{3}}{2}$	-1	$-\dfrac{\sqrt{3}}{2}$	$-\dfrac{\sqrt{2}}{2}$	$-\dfrac{1}{2}$	0	$\dfrac{1}{2}$	$\dfrac{\sqrt{2}}{2}$	$\dfrac{\sqrt{3}}{2}$	1

If we plot all of these points and connect them into a smooth curve, we get:

Figure 3

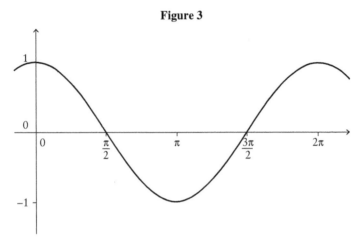

Here are the important things to remember about a cosine graph:

- Its minimum y-value is -1 and its maximum y-value is 1.
- The period of one wavelength is 2π.
- We refer to the *beginning* of the graph as the point $(0, 1)$.
- The midline of the graph is the x-axis.

Notice that it has the same shape as the sine graph, which should make sense because we learned that $\cos x = \sin(90° - x)$. Thus, the graphs of the two functions are the same, merely shifted by $90°$ or $\frac{\pi}{2}$. Sine and cosine graphs are often referred to as *sinusoidal* functions, because they are all based on the sine function.

We are going to now learn how to graph more complicated sine and cosine graphs. First, let's learn some terminology.

- The *amplitude* of a sine or cosine graph is the distance from the midline to either the maximum or minimum of the graph. We can also find the amplitude by finding $\frac{|\max - \min|}{2}$.

For a basic sine or cosine graph, the amplitude is 1.

- The *period* of a sine or cosine graph is the distance along the x-axis of one wavelength.

For a basic sine or cosine graph, the period is 2π.

- The distance from the y-axis to the starting point of the graph is called the *horizontal shift*. This is sometimes also called the *phase shift*.

For example, in the figure below, the sine curve "begins" on the x-axis at $\left(0, \frac{\pi}{6}\right)$, so the horizontal shift is $\frac{\pi}{6}$ units to the right. Notice how all of the x-coordinates have $\frac{\pi}{6}$ added to them.

Figure 4

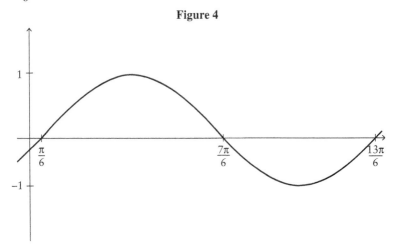

Now we will look at another shift:

- The distance from the x-axis that the midline is shifted is called the *vertical shift*.

For example, in the figure below, the sine curve has its midline at $y = 1$. Notice how it now has its maximum at $\left(\frac{\pi}{2}, 2\right)$ and its minimum at $\left(\frac{3\pi}{2}, 0\right)$.

Figure 5

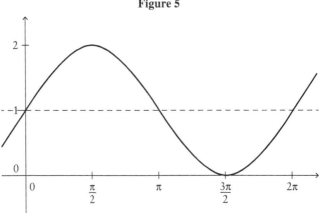

All of these terms refer to the graphs of any trig function. Now we will learn how to transform a sine or a cosine graph. Later, we will learn how to graph the other four trig functions.

Amplitude

> If we have an equation of the form $y = A\sin x$ or $y = A\cos x$, then the amplitude of the graph will be $|A|$.

For example, the graph $y = 2\sin x$ has an amplitude of 2. This means that the maximum of the graph will be $\left(\dfrac{\pi}{2}, 2\right)$ and the minimum will be $\left(\dfrac{3\pi}{2}, -2\right)$, and that all of the y-values of the graph will be multiplied by 2. Otherwise, it will look exactly like the graph of $y = \sin x$. In the figure below, we graph both $y = \sin x$ and $y = 2\sin x$ so that you can see the difference:

Figure 6

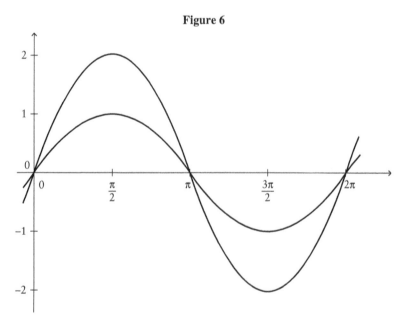

Notice how the graph still has its maximum at $\dfrac{\pi}{2}$ and its minimum at $\dfrac{3\pi}{2}$, and that it still crosses the x-axis at 0, π, and 2π.

Let's do another one. Below is the graph of one period of $y = 3\cos x$:

UNIT SEVEN: Sine and Cosine Graphs

Figure 7

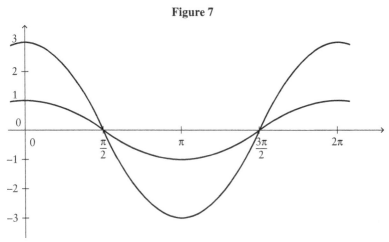

Again, notice that the only difference between the two graphs is the maximum and the minimum.

Of course, A doesn't have to be an integer. Let's graph one period of $y = \frac{1}{2}\sin x$:

Figure 8

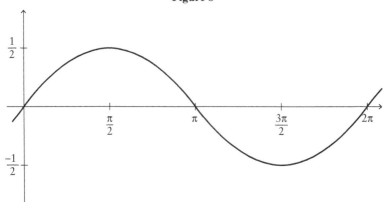

Now let's learn how to graph different length periods.

Period

> If we have an equation of the form $y = \sin(Bx)$ or $y = \cos(Bx)$, then the period of the graph will be $\frac{2\pi}{B}$.

For example, the graph $y = \sin(2x)$ has a period of $\frac{2\pi}{2} = \pi$. This means that the graph of one wavelength will be a distance of π, rather than 2π, and that all of the x-coordinates of the graph will be divided by 2. The graph is below:

Figure 9

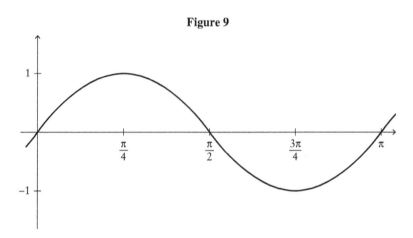

Notice that the amplitude is unchanged, and just the period is changed.

Let's do another one:

Below is the graph of one period of $y = \cos\left(\frac{x}{2}\right)$:

Figure 10

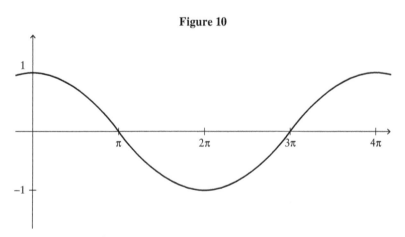

Here, the period is $\frac{2\pi}{1/2} = 4\pi$. Thus, each of the x-coordinates will be divided by $\frac{1}{2}$ (in other words, multiplied by 2).

UNIT SEVEN: Sine and Cosine Graphs

Now let's do a graph where we change both the amplitude and the period. For example, let's graph one period of $y = 3\sin(4x)$. The amplitude will be 3 and the period will be $\dfrac{2\pi}{4} = \dfrac{\pi}{2}$:

Figure 11

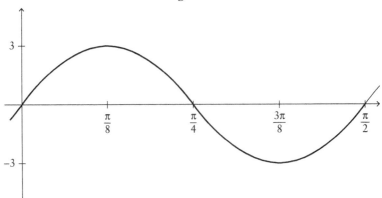

Notice that the maximum is at $\left(\dfrac{\pi}{8}, 3\right)$ and the minimum is at $\left(\dfrac{3\pi}{8}, -3\right)$, and that the graph crosses the x-axis at 0, $\dfrac{\pi}{4}$, and $\dfrac{\pi}{2}$.

Let's do another example. Graph one period of $y = \dfrac{1}{2}\cos\left(\dfrac{\pi}{2}x\right)$. Now the amplitude is $\dfrac{1}{2}$ and the period is $\dfrac{2\pi}{\pi/2} = 4$.

Figure 12

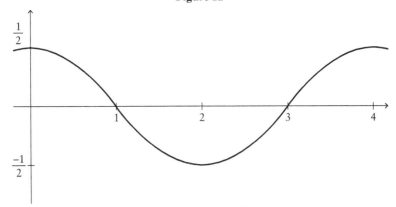

Notice that the maximum is at $\left(\left(0, \dfrac{1}{2}\right)\text{ and }\left(4, \dfrac{1}{2}\right)\right)$ and the minimum is at $\left(2, -\dfrac{1}{2}\right)$.

Now let's learn how to graph a horizontal shift.

Horizontal Shift

> If we have an equation of the form $y = \sin(x \pm C)$ or $y = \cos(x \pm C)$, then we shift the curve C units to the right if C is subtracted from x, and we shift the curve C units to the left if C is added to x.

For example, suppose we want to graph one period of $y = \cos\left(x - \dfrac{\pi}{3}\right)$. We will add $\dfrac{\pi}{3}$ units to each of the x-coordinates of the graph of $y = \cos x$.

Figure 13

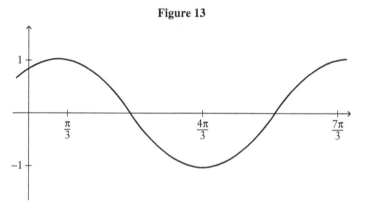

Notice that the maximum is still 1, but that it now occurs at $\left(\dfrac{\pi}{3}, 1\right)$, instead of at $(0, 1)$. Similarly, the minimum is still -1 but it now occurs at $\left(\dfrac{4\pi}{3}, -1\right)$, instead of at $(\pi, -1)$.

Let's do another one. Graph one period of $y = \sin\left(x + \dfrac{\pi}{4}\right)$. Now we will subtract $\dfrac{\pi}{4}$ units from each of the x-coordinates of the graph $y = \sin x$.

Figure 14

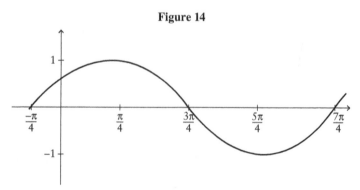

UNIT SEVEN: Sine and Cosine Graphs

Notice that the maximum now is at $\left(\frac{\pi}{4}, 1\right)$, instead of $\left(\frac{\pi}{2}, 1\right)$, and that the minimum is at $\left(\frac{5\pi}{4}, -1\right)$, instead of $\left(\frac{3\pi}{2}, -1\right)$.

Now we will do a graph with the three transformations that we have learned so far. Let's graph one period of $y = 3\sin 2\left(x - \frac{\pi}{6}\right)$. Here, the amplitude is 3, the period is $\frac{2\pi}{2} = \pi$, and the horizontal shift is $\frac{\pi}{6}$ units to the right.

Figure 15

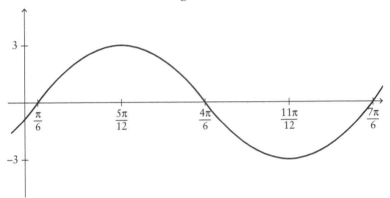

Now the maximum is at $\left(\frac{5\pi}{12}, 3\right)$, and the minimum is at $\left(\frac{11\pi}{12}, -3\right)$. How did we get those values? Remember that the period is now π and that the curve "begins" at $\frac{\pi}{6}$. This means that the first wavelength will "end" at $\frac{\pi}{6} + \pi = \frac{7\pi}{6}$. The midpoint of $\frac{\pi}{6}$ and $\frac{7\pi}{6}$ is $\frac{4\pi}{6}$. From there, we can find the rest of the *x*-coordinates by finding the midpoint of $\frac{\pi}{6}$ and $\frac{4\pi}{6}$, and so on. The *y*-coordinates are found by multiplying 1 and −1 by 3.

Now let's do the final transformation, the vertical shift.

Vertical Shift

If we have an equation of the form $y = \sin x \pm D$ or $y = \cos x \pm D$, then we shift the curve *D* units up if *D* is added to $\sin x$, and we shift the curve *D* units down if *D* is subtracted from $\sin x$.

Notice that *D* is **not** added or subtracted to *x*. Sometimes, you will see the equation written as $y = \pm D + \sin x$ to avoid confusion. Let's do an example. Graph one period of $y = \sin x + 2$. All that we do is shift the graph of $\sin x$ up 2 units from the *x*-axis. You will find that it is helpful to draw a dotted midline to help keep track of the shift.

Figure 16

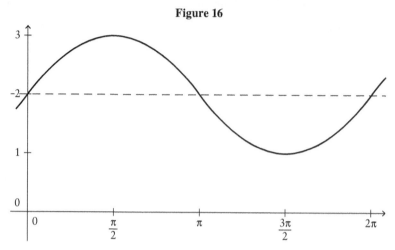

Notice that the *x*-coordinates did not change from the original graph $y = \sin x$. Only the *y*-coordinates changed.

Let's do another example. Graph one period of $y = \cos x - 1$.

Figure 17

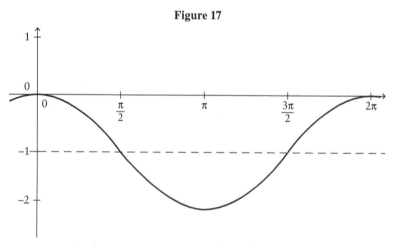

Again, all that we had to do is shift the cosine curve down 1 unit from its original position.

UNIT SEVEN: Sine and Cosine Graphs   81

Are you ready to do all of the transformations at once? Here we go!

Let's graph one period of $y = 2\sin 3\left(x - \dfrac{\pi}{4}\right) + 1$. What are the transformations?

Amplitude is 2.

Period is $\dfrac{2\pi}{3}$.

Horizontal Shift is $\dfrac{\pi}{4}$ units to the right.

Vertical Shift is up 1 unit.

Figure 18

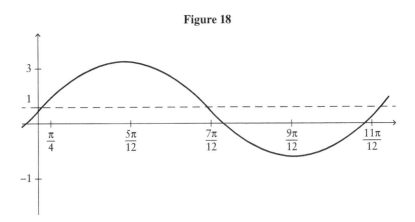

Here, in order to find the x-coordinates, we took the starting point, $\dfrac{\pi}{4}$, and added the period $\dfrac{2\pi}{3}$ to get $\dfrac{\pi}{4} + \dfrac{2\pi}{3} = \dfrac{11\pi}{12}$ as the endpoint of one period. From there, we cut the interval between $\dfrac{\pi}{4}$ and $\dfrac{11\pi}{12}$ into quarters. In order to find the y-coordinates, we first multiplied 1 and -1 by the amplitude, 2 to get the values 2 and -2, respectively. Then we added the vertical shift of 1 to each of those to get 3 and -1, respectively.

Let's do another one:

Graph one period of $y = 4\cos\dfrac{\pi}{3}(x+1) - 2$. What are the transformations?

Amplitude is 4.

Period is $\dfrac{2\pi}{\pi/3} = 6$.

Horizontal Shift is 1 unit to the left.

Vertical Shift is down 2 units.

Figure 19

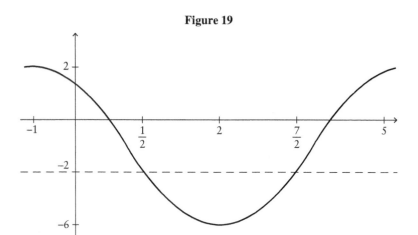

Here, in order to find the *x*-coordinates, we took the starting point, −1, and added the period 6 to get 5 as the endpoint. From there, we cut the interval between −1 and 5 into quarters. In order to find the *y*-coordinates, we first multiplied 1 and −1 by the amplitude, 4, to get 4 and −4, respectively. Then we subtracted the vertical shift of 2 to get 2 and −6, respectively.

One final type of transformation: what if the amplitude is a negative number? Simple. You just turn the graph upside down. For example, let's graph one period of $y = -3\sin(4x)$.

Figure 20

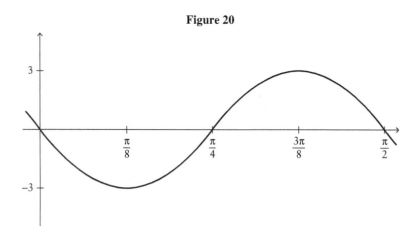

Here the amplitude is 3 and the period is $\frac{2\pi}{4} = \frac{\pi}{2}$. Then we flip the graph.

Now let's do some practice problems.

UNIT SEVEN: Sine and Cosine Graphs

Practice Problems

Practice problem 1: Graph one period of $y = 5\cos(2x)$.

Practice problem 2: Graph one period of $y = \frac{1}{4}\sin\left(\frac{x}{3}\right)$.

Practice problem 3: Graph one period of $y = \sin\left(x + \frac{\pi}{6}\right) - 1$.

Practice problem 4: Graph one period of $y = \cos\left(x - \frac{\pi}{6}\right) - 3$.

Practice problem 5: Graph one period of $y = 2\sin\frac{\pi}{4}(x - 3) + 1$.

Practice problem 6: Graph one period of $y = 4\cos 2\left(x + \frac{\pi}{3}\right) - 3$.

Practice problem 7: Graph one period of $y = \frac{2}{3}\sin\frac{1}{2}\left(x + \frac{\pi}{4}\right) + 1$.

Practice problem 8: Graph one period of $y = \frac{3}{5}\cos\frac{\pi}{6}(x + 1) - \frac{2}{5}$.

Practice problem 9: Graph one period of $y = -2\sin 3\left(x + \frac{\pi}{6}\right) + 5$.

Practice problem 10: Graph one period of $y = -3\cos\frac{\pi}{3}(x - 2) - 1$.

Solutions to the Practice Problems

Solution to practice problem 1: *Graph one period of $y = 5\cos(2x)$.*

The transformations are:
Amplitude: 5
Period: $\frac{2\pi}{2} = \pi$
There are no horizontal or vertical shifts.
The graph looks like this:

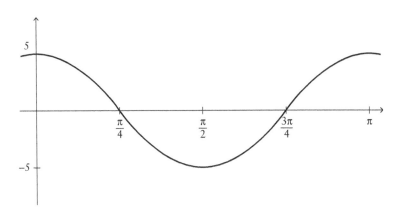

Solution to practice problem 2: *Graph one period of* $y = \frac{1}{4}\sin\left(\frac{x}{3}\right)$.
The transformations are:
Amplitude: $\frac{1}{4}$
Period: $\frac{2\pi}{1/3} = 6\pi$
There are no horizontal or vertical shifts.
The graph looks like this:

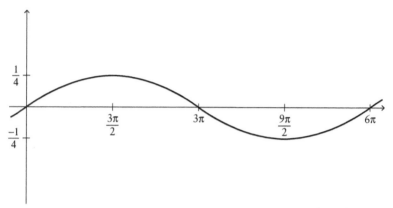

Solution to practice problem 3: *Graph one period of* $y = \sin\left(x + \frac{\pi}{6}\right) - 1$.
The transformations are:
Amplitude: 1
Period: 2π
Horizontal Shift: $\frac{\pi}{6}$ units to the left
Vertical Shift: 1 unit down
The graph looks like this:

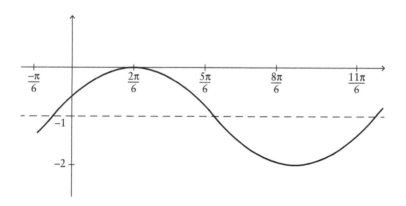

Solution to practice problem 4: *Graph one period of* $y = \cos\left(x - \dfrac{\pi}{6}\right) - 3$.

The transformations are:
Amplitude: 1
Period: 2π
Horizontal Shift: $\dfrac{\pi}{6}$ units to the right
Vertical Shift: 3 units down

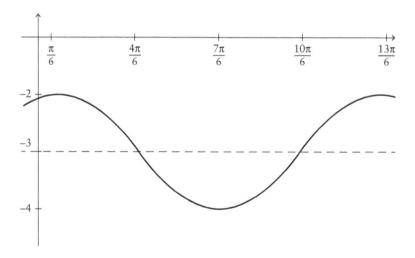

Solution to practice problem 5: *Graph one period of* $y = 2\sin\dfrac{\pi}{4}(x - 3) + 1$.

The transformations are:
Amplitude: 2
Period: $\dfrac{2\pi}{\pi/4} = 8$
Horizontal Shift: 3 units to the right
Vertical Shift: 1 unit up

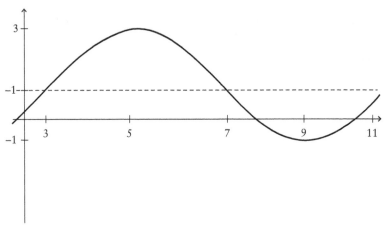

Solution to practice problem 6: *Graph one period of* $y = 4\cos 2\left(x + \dfrac{\pi}{3}\right) - 3$.

The transformations are:
Amplitude: 4
Period: $\dfrac{2\pi}{2} = \pi$
Horizontal Shift: $\dfrac{\pi}{3}$ units to the left
Vertical Shift: 3 units down

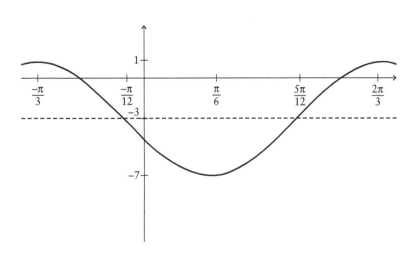

Solution to practice problem 7: *Graph one period of* $y = \frac{2}{3}\sin\frac{1}{2}\left(x+\frac{\pi}{4}\right)+1$.
The transformations are:

Amplitude: $\frac{2}{3}$

Period: $\frac{2\pi}{1/2} = 4\pi$

Horizontal Shift: $\frac{\pi}{4}$ units to the left

Vertical Shift: 1 unit up

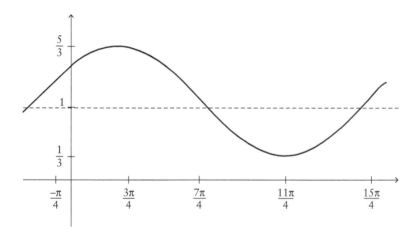

Solution to practice problem 8: *Graph one period of* $y = \frac{3}{5}\cos\frac{\pi}{6}(x-1)-\frac{2}{5}$.
The transformations are:

Amplitude: $\frac{3}{5}$

Period: $\frac{2\pi}{\pi/6} = 12$

Horizontal Shift: 1 unit to the right

Vertical Shift: $\frac{2}{5}$ unit down

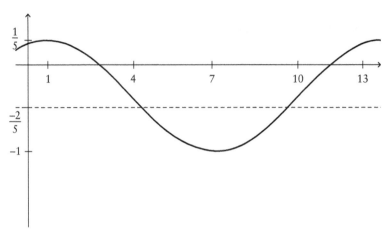

Solution to practice problem 9: *Graph one period of* $y = -2\sin\frac{2}{3}\left(x + \frac{\pi}{6}\right) + 5$

The transformations are:
Amplitude: 2
Period: $\frac{2\pi}{2/3} = 3\pi$
Horizontal Shift: $\frac{\pi}{6}$ units to the left
Vertical Shift: 5 units up
The graph is flipped upside down.

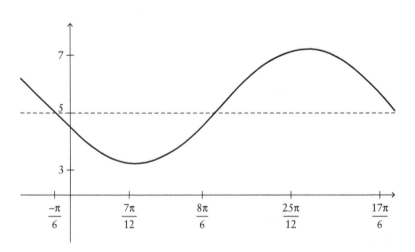

Solution to practice problem 10: *Graph one period of* $y = -3\cos\frac{\pi}{3}(x-2) - 1$

The transformations are:
Amplitude: 3
Period: $\frac{2\pi}{\pi/3} = 6$
Horizontal Shift: 2 units to the right
Vertical Shift: 1 unit down
The graph is flipped upside down.

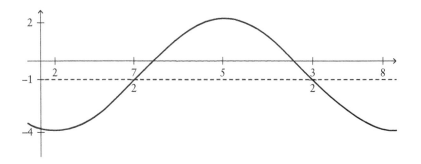

UNIT EIGHT

Graphing Tangent, Cotangent, Secant, and Cosecant

Now we are going to learn how to graph the tangent function. First, let's make a table of the value of $\tan x$ as x goes from 0 to 2π radians.

x	0	$\frac{\pi}{6}$	$\frac{\pi}{4}$	$\frac{\pi}{3}$	$\frac{\pi}{2}$	$\frac{2\pi}{3}$	$\frac{3\pi}{4}$	$\frac{5\pi}{6}$	π	$\frac{7\pi}{6}$	$\frac{5\pi}{4}$	$\frac{4\pi}{3}$	$\frac{3\pi}{2}$	$\frac{5\pi}{3}$	$\frac{7\pi}{4}$	$\frac{11\pi}{6}$	2π
$\tan x$	0	$\frac{1}{\sqrt{3}}$	1	$\sqrt{3}$	–	$-\frac{1}{\sqrt{3}}$	-1	$-\sqrt{3}$	0	$\frac{1}{\sqrt{3}}$	1	$\sqrt{3}$	–	$-\frac{1}{\sqrt{3}}$	-1	$-\sqrt{3}$	0

There are a couple of differences between the tangent graph and the sine and cosine graphs that we looked at previously. First, notice that the period of $\tan x$ is π, not 2π. Second, notice that tangent is undefined at $\frac{\pi}{2}$ and $\frac{3\pi}{2}$ (and will be for all of the odd multiples of $\frac{\pi}{2}$), so we have vertical asymptotes at those values of x.

Let's graph two periods of $y = \tan x$:

Figure 1

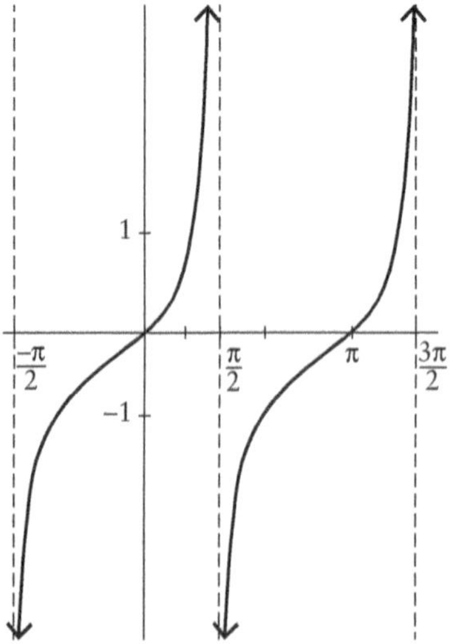

UNIT EIGHT: Graphing Tangent, Cotangent, Secant, and Cosecant

Notice that we graphed the function on the interval from $-\frac{\pi}{2}$ to $\frac{3\pi}{2}$, instead of the interval from 0 to 2π. This is purely for aesthetic reasons. The tangent graph can be transformed in much the same way that sine and cosine graphs can.

Let's graph one period of $y = \tan 2x$:

Figure 2

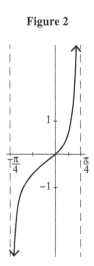

Here the period is $\frac{\pi}{2}$ (remember that the period of $\tan x$ is π!).

Now let's graph one period of $y = \tan\left(\frac{x}{2}\right) + 1$:

Figure 3

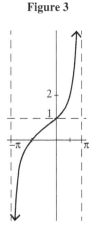

Here the period is $\frac{\pi}{1/2} = 2\pi$ and the vertical shift is up 1 unit.

Note: The other three functions – cosecant, secant, and cotangent – are the reciprocals of sine, cosine, and tangent, respectively. Because of this, almost all trig applications can be done with the basic functions and don't really need the reciprocal functions. This is probably why you don't have buttons for them on your calculators! But, you should be familiar with the three graphs. Of course, they can be transformed just as sine and cosine can.

Now let's graph two periods of $y = \cot x$:

Figure 4

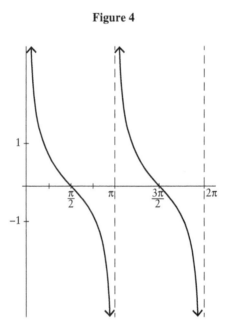

As with tangent, the period is π. Also, because cotangent is the reciprocal of tangent, it is undefined at 0, π and 2π (and for all of the multiples of π because tangent is 0 there), so we have vertical asymptotes at those values of x.

Now let's look at cosecant. Remember that cosecant is the reciprocal of sine. This means that it is undefined wherever sine is 0; namely at 0, π and 2π (and for of the multiples of π). Thus, we have vertical asymptotes at those values of x.

Here is one period of $y = \csc x$:

UNIT EIGHT: Graphing Tangent, Cotangent, Secant, and Cosecant   93

Figure 5

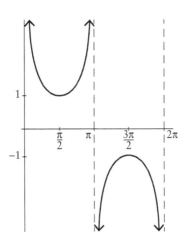

The period is 2π, just as it was with sine.

Finally, let's look at secant. Remember that secant is the reciprocal of cosine. This means that it is undefined where cosine is 0, namely at $\frac{\pi}{2}$ and $\frac{3\pi}{2}$ (and for all of the odd multiples of $\frac{\pi}{2}$). Thus, we have vertical asymptotes at those values of x.

Here is one period of $y = \sec x$:

Figure 6

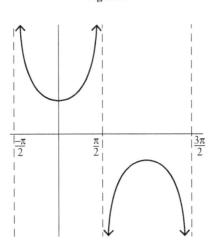

Again, the period is 2π, just as it was with cosine. As with tangent, we graphed the function on the interval from $-\frac{\pi}{2}$ to $\frac{3\pi}{2}$, instead of the interval from 0 to 2π. This is purely for aesthetic reasons.

Let's do a few examples.

Example 1: Graph one period of $y = \cot\left(\frac{x}{2}\right) + 1$.

Figure 7

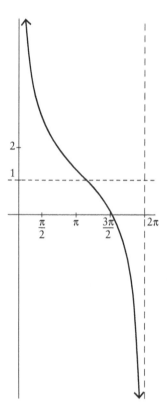

Here, the period is $\frac{\pi}{1/2} = 2\pi$, and the vertical shift is 1 unit up.

Example 2: Graph one period of $y = \csc\left(\frac{\pi}{2}x\right) - 1$.

UNIT EIGHT: Graphing Tangent, Cotangent, Secant, and Cosecant   95

Figure 8

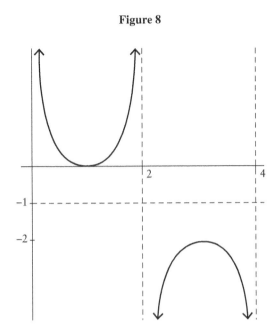

Here the period is $\dfrac{2\pi}{\pi/2} = 4$, and the vertical shift is 1 unit down.

Example 3: Graph one period of $y = \sec(3x) + 2$.

Figure 9

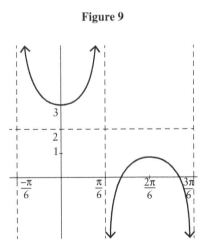

Here the period is $\frac{2\pi}{3}$, and the vertical shift is 2 units up.

Although we could give tangent, cotangent, secant, and cosecant as full a treatment as we did for the sine and cosine graphs, they are used much less often in applications of science and engineering. Thus you will see that your teachers will give them less attention and sometimes skip them altogether. Just in case, you should know what they look like and how to graph them.

Time to do some practice problems!

Practice Problems

Practice problem 1: Graph one period of $y = \tan\left(\frac{x}{3}\right)$.

Practice problem 2: Graph one period of $y = \tan\left(x - \frac{\pi}{6}\right) + 2$.

Practice problem 3: Graph one period of $y = \cot\left(x + \frac{\pi}{4}\right) - 1$.

Practice problem 4: Graph one period of $y = \csc\left(x - \frac{\pi}{3}\right) + 1$.

Practice problem 5: Graph one period of $y = \sec\frac{\pi}{4}(x - 1) + 2$.

Practice problem 6: Graph one period of $y = \csc\left(x + \frac{\pi}{3}\right) - 1$.

Solutions to the Practice Problems

Solution to practice problem 1: *Graph one period of* $y = \tan\left(\frac{x}{3}\right)$.

The only transformation is:

Period: $\frac{\pi}{1/3} = 3\pi$

UNIT EIGHT: Graphing Tangent, Cotangent, Secant, and Cosecant   97

The graph looks like this:

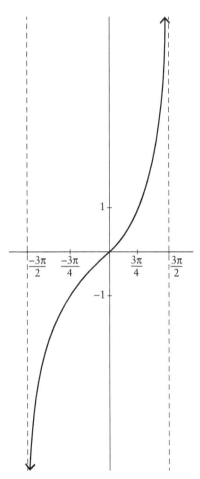

Solution to practice problem 2: *Graph one period of* $y = \tan\left(x - \dfrac{\pi}{6}\right) + 2$.

The transformations are:

Horizontal shift: $\dfrac{\pi}{6}$ units to the right
Vertical shift: 2 units up

The graph looks like this:

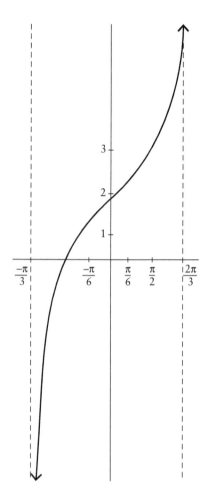

Solution to practice problem 3: *Graph one period of* $y = \cot\left(x + \dfrac{\pi}{4}\right) - 1.$

The transformations are:

Horizontal shift: $\dfrac{\pi}{4}$ units to the left
Vertical Shift: 1 unit down

The graph looks like this:

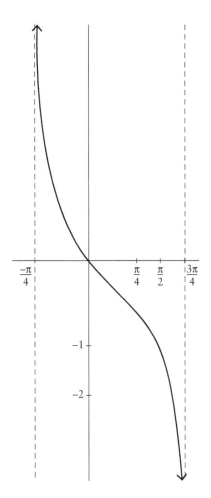

Solution to practice problem 4: *Graph one period of* $y = \csc\left(x - \dfrac{\pi}{3}\right) + 1.$

The transformations are:

Horizontal Shift: $\dfrac{\pi}{3}$ units to the right

Vertical Shift: 1 unit up

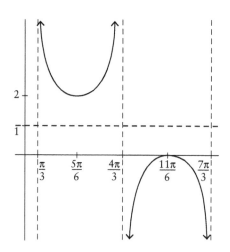

Solution to practice problem 5: *Graph one period of* $y = \sec\dfrac{\pi}{4}(x-1)+2.$

The transformations are:

Period: $\dfrac{2\pi}{\pi/4} = 8$

Horizontal Shift: 1 units to the right
Vertical Shift: 2 units up

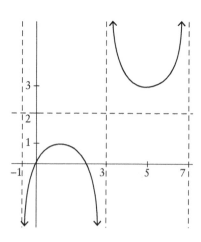

Solution to practice problem 6: *Graph one period of* $y = \csc\left(x + \dfrac{\pi}{3}\right) - 1$.
Horizontal Shift: $\dfrac{\pi}{3}$ units to the left
Vertical Shift: 1 unit down

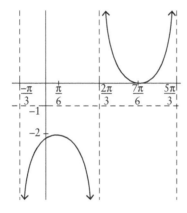

UNIT NINE

Inverse Trigonometric Functions

So far, we have worked with the six basic trig functions. When we are given an angle, we find one or more trigonometric values for that angle. Now we will learn how to go in the other direction. That is, given the value of a trig function of an angle, what is the angle? There is one problem. We know, for example, that $\sin\frac{\pi}{6} = \frac{1}{2}$. But, if we are told that $\sin x = \frac{1}{2}$, x could be $\frac{\pi}{6}$, or $\frac{5\pi}{6}$, or $\frac{13\pi}{6}$. In fact, there are an infinite number of possibilities for x. So, we need to keep the trig functions *one-to-one*. In other words, to make sure that there is only one value of x, we have to restrict the domain of the inverse function.

Here's what we do:

If we restrict the domain of $y = \sin x$ to where $-\frac{\pi}{2} \leq x \leq \frac{\pi}{2}$, then the function will be 1-1. If you look at the graphs below, you can see that $y = \sin x$ and $y = \sin^{-1} x$, will both pass the vertical line tests. That is, you will get a unique answer for the inverse sine of an angle.

Figure 1

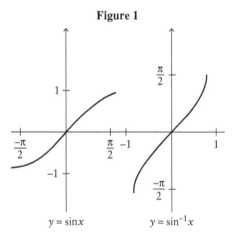

$y = \sin x$ $y = \sin^{-1} x$

This means that the domain of $y = \sin^{-1} x$ is $-1 \leq x \leq 1$, and the range will then be $-\frac{\pi}{2} \leq y \leq \frac{\pi}{2}$, which is what we want.

Notice that notation that we used for inverse sine: $\sin^{-1} x$. Do **not** confuse this with $\frac{1}{\sin x}$, which is the reciprocal function $\csc x$. To avoid this confusion, many people call inverse sine *arcsine* (arcsin x) instead. (Your calculator uses the first notation mainly to save space.) We will use the two notations interchangeably so that you will get used to seeing both of them.

Example 1: Let's evaluate $y = \sin^{-1}\left(\dfrac{\sqrt{3}}{2}\right)$.

We know that $\sin\dfrac{\pi}{3} = \dfrac{\sqrt{3}}{2}$, so $\sin^{-1}\left(\dfrac{\sqrt{3}}{2}\right) = \dfrac{\pi}{3}$. If we hadn't restricted the domain, we would have an infinite number of answers.

Now let's evaluate $y = \arcsin\left(-\dfrac{\sqrt{3}}{2}\right)$. We know that $\sin\left(-\dfrac{\pi}{3}\right) = -\dfrac{\sqrt{3}}{2}$, so $\arcsin\left(-\dfrac{\sqrt{3}}{2}\right) = -\dfrac{\pi}{3}$.

Now let's do inverse cosine. Just as with inverse sine, we restrict $y = \cos^{-1} x$ so that the function is 1-1. Here the domain of $y = \cos^{-1} x$ is $-1 \leq x \leq 1$ and the range is $0 \leq y \leq \pi$. Let's look at the graphs so that you can see why:

Figure 2

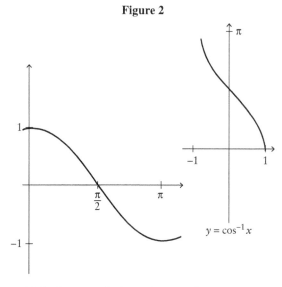

Another way to think about $y = \sin^{-1} x$ and $y = \cos^{-1} x$ is to look at the unit circle. The y-values of inverse sine of x will always be found on the right-hand side of the unit circle. The y-values of inverse cosine of x will always be found on the top half of the unit circle.

Figure 3

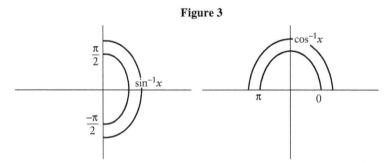

What about the inverse tangent function? Let's look at the graphs:

Figure 4

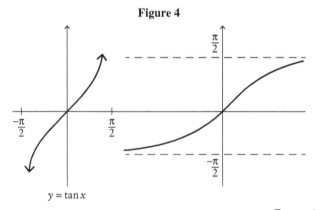

$y = \tan x$

Here the domain for $y = \tan^{-1} x$ is $-\infty < x < \infty$ and the range is $-\frac{\pi}{2} < y < \frac{\pi}{2}$. Notice that it is almost the same range as $\sin^{-1} x$, except that it doesn't include the endpoints. As with $y = \sin^{-1} x$, the y-values of $y = \tan^{-1} x$ will always be found on the right-hand side of the unit circle.

Let's do some more examples.

Example 2: Let's find $\cos^{-1}\left(\frac{1}{2}\right)$.

We know that $\cos\left(\frac{\pi}{3}\right) = \frac{1}{2}$, so $\cos^{-1}\left(\frac{1}{2}\right) = \frac{\pi}{3}$.

Let's find $\arctan(-1)$. We know that $\tan\left(-\frac{\pi}{4}\right) = -1$, so $\arctan(-1) = \frac{\pi}{4}$.

What about the inverse functions of cotangent, cosecant, and secant? We don't really need them because we can always work with their reciprocal functions

UNIT NINE: Inverse Trigonometric Functions

(sine, cosine, and tangent). Furthermore, restricting the range of arcsec and arccsc is trickier than with arcsine and arccos. Therefore, we won't discuss them in this book.

What happens when we find the composition of a trig function and its inverse? The following are true wherever they are defined: $\sin(\sin^{-1} x) = x$, $\cos(\cos^{-1} x) = x$, and $\tan(\tan^{-1} x) = x$.

However, the following examples are only true because we restricted the domain of sine, cosine, and tangent.

Example 3: Let's find $\sin\left(\sin^{-1}\dfrac{1}{2}\right)$.

We know that $\sin^{-1}\dfrac{1}{2} = \dfrac{\pi}{6}$ and we know that $\sin\dfrac{\pi}{6} = \dfrac{1}{2}$, so $\sin\left(\sin^{-1}\dfrac{1}{2}\right) = \dfrac{1}{2}$.

For another example, let's find $\cos^{-1}\left(\cos\dfrac{\pi}{4}\right)$. We know that $\cos\dfrac{\pi}{4} = \dfrac{\sqrt{2}}{2}$, and we know that $\cos^{-1}\dfrac{\sqrt{2}}{2} = \dfrac{\pi}{4}$, so $\cos^{-1}\left(\cos\dfrac{\pi}{4}\right) = \dfrac{\pi}{4}$.

These aren't very interesting, but what about $\tan(\sin^{-1} x)$? Remember the definitions of the trig functions? Well, $\sin^{-1} x$ means that we can construct a triangle with an angle θ, where $\sin\theta = x$ whose sine is x, as in the figure below:

Figure 5

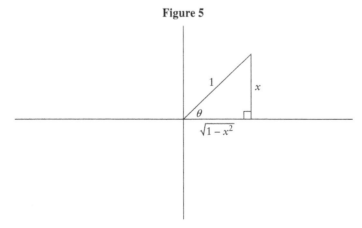

Now we can find the other leg from the Pythagorean Theorem. We get that the other leg is $\sqrt{1-x^2}$. Now we just have to find $\tan\theta$, which is $\tan\theta = \dfrac{x}{\sqrt{1-x^2}}$.

Notice a couple of things about this example. First, the inverse trig function of a number gives us an angle. Second, in this example we assumed that x was a positive number. If x had been a negative number, the triangle would have been located in Quadrant IV, which means that tangent would have been negative. Keep in mind that we need to know which quadrant we are working with when solving these kinds of problems.

Let's do some more examples.

Example 4: Evaluate $\tan\left(\cos^{-1}\dfrac{4}{5}\right)$.

Because $\dfrac{4}{5}$ is positive, the angle must be located in Quadrant I, and the tangent will be positive. Let's draw a triangle with an angle θ, where $\cos\theta = \dfrac{4}{5}$:

Figure 6

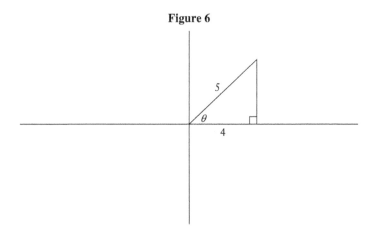

Now we can find the other leg of the triangle using the Pythagorean Theorem. You should get $\sqrt{5^2 - 4^2} = 3$. This means that $\tan\theta = \dfrac{3}{4}$. Therefore, $\tan\left(\cos^{-1}\dfrac{4}{5}\right) = \dfrac{3}{4}$.

Example 5: Evaluate $\sin\left(\tan^{-1}\left(-\dfrac{2}{7}\right)\right)$.

Because $-\dfrac{2}{7}$ is negative, the angle must be located in Quadrant IV, and that the sine will be negative. Let's draw a triangle, with an angle θ, where $\tan\theta = -\dfrac{2}{7}$:

UNIT NINE: Inverse Trigonometric Functions

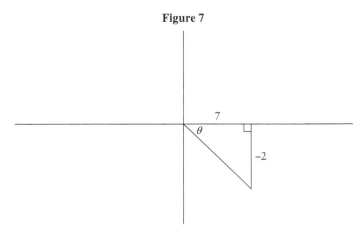

Figure 7

Now we can find the hypotenuse of the triangle using the Pythagorean Theorem. You should get $\sqrt{2^2 + 7^2} = \sqrt{53}$. This means that $\sin\theta = -\frac{2}{\sqrt{53}}$. Therefore, $\sin\left(\tan^{-1}\left(-\frac{2}{7}\right)\right) = -\frac{2}{\sqrt{53}}$.

How about one where the composition is the other way?

Example 6: Evaluate $\cos^{-1}\left(\sin\frac{5\pi}{6}\right)$.

We know that $\sin\left(\frac{5\pi}{6}\right) = \frac{1}{2}$. Now we just have to find $\cos^{-1}\frac{1}{2} = \frac{\pi}{3}$. Notice something important: even though $\frac{5\pi}{6}$ is in Quadrant II, the inverse cosine of $\frac{1}{2}$ is located in Quadrant I. This is because we strictly defined the domain and range of the functions to avoid multiple answers.

Time to practice!

Practice Problems

Practice problem 1: Find the value of $\sin^{-1}\left(-\frac{\sqrt{2}}{2}\right)$.

Practice problem 2: Find the value of $\cos^{-1}\left(\frac{\sqrt{3}}{2}\right)$.

Practice problem 3: Find the value of $\arctan\left(\sqrt{3}\right)$.

Practice problem 4: Find the value of $\sin^{-1}(0)$.

Practice problem 5: Find the value of $\cos^{-1}(-1)$.

Practice problem 6: Find the value of $\cos\left(\arctan\dfrac{1}{\sqrt{3}}\right)$.

Practice problem 7: Find the value of $\arcsin\left(\cos\dfrac{\pi}{3}\right)$.

Practice problem 8: Find the value of $\sin\left(\tan^{-1}\left(\dfrac{3}{8}\right)\right)$.

Practice problem 9: Find the value of $\tan\left(\arccos\left(-\dfrac{6}{7}\right)\right)$.

Practice problem 10: Find the value of $\cos^{-1}\left(\cos\left(\dfrac{7\pi}{4}\right)\right)$.

Practice problem 11: Find the value of $\arcsin\left(\sin\left(-\dfrac{4\pi}{3}\right)\right)$.

Practice problem 12: Find the value of $\sin\left(\tan^{-1}\dfrac{x}{2}\right)$.

Solutions to the practice problems

Solution to practice problem 1: *Find the value of* $\sin^{-1}\left(-\dfrac{\sqrt{2}}{2}\right)$.

We know that $\sin\left(-\dfrac{\pi}{4}\right) = -\dfrac{\sqrt{2}}{2}$, so $\sin^{-1}\left(-\dfrac{\sqrt{2}}{2}\right) = -\dfrac{\pi}{4}$.

Solution to practice problem 2: *Find the value of* $\cos^{-1}\left(\dfrac{\sqrt{3}}{2}\right)$.

We know that $\cos\left(\dfrac{\pi}{6}\right) = \dfrac{\sqrt{3}}{2}$, so $\cos^{-1}\left(\dfrac{\sqrt{3}}{2}\right) = \dfrac{\pi}{6}$.

Solution to practice problem 3: *Find the value of* $\arctan(\sqrt{3})$.

We know that $\tan\left(\dfrac{\pi}{3}\right) = \sqrt{3}$, so $\tan^{-1}(\sqrt{3}) = \dfrac{\pi}{3}$.

Solution to practice problem 4: *Find the value of* $\sin^{-1}(0)$.
 We know that $\sin 0 = 0$, so $\sin^{-1}(0) = 0$.

UNIT NINE: Inverse Trigonometric Functions

Solution to practice problem 5: *Find the value of* $\cos^{-1}(-1)$.
We know that $\cos \pi = -1$, so $\cos^{-1}(-1) = \pi$.

Solution to practice problem 6: *Find the value of* $\cos\left(\arctan\dfrac{1}{\sqrt{3}}\right)$.
We know that $\tan\dfrac{\pi}{6} = \dfrac{1}{\sqrt{3}}$, so $\tan^{-1}\dfrac{1}{\sqrt{3}} = \dfrac{\pi}{6}$. Now, we just have to evaluate $\cos\dfrac{\pi}{6} = \dfrac{\sqrt{3}}{2}$. Therefore, $\cos\left(\arctan\dfrac{1}{\sqrt{3}}\right) = \dfrac{\sqrt{3}}{2}$.

Solution to practice problem 7: *Find the value of* $\arcsin\left(\cos\left(\dfrac{\pi}{3}\right)\right)$.
We know that $\cos\dfrac{\pi}{3} = \dfrac{1}{2}$. Now we just have to evaluate $\arcsin\dfrac{1}{2}$.
Because $\sin\dfrac{\pi}{6} = \dfrac{1}{2}$, we get $\arcsin\dfrac{1}{2} = \dfrac{\pi}{6}$. Therefore, $\arcsin\left(\cos\left(\dfrac{\pi}{3}\right)\right) = \dfrac{\pi}{6}$.

Solution to practice problem 8: *Find the value of* $\sin\left(\tan^{-1}\left(\dfrac{3}{8}\right)\right)$.

This problem tells us that we have an angle whose tangent is $\dfrac{3}{8}$, and we need to find the sine of that angle. Let's draw a picture:

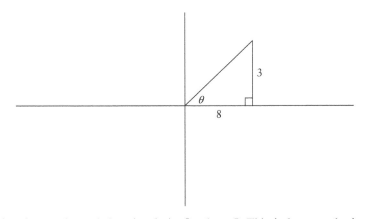

Notice that we located the triangle in Quadrant I. This is because the inverse tangent of a positive number is an angle in Quadrant I. Remember that the sine of an angle in Quadrant I will be positive. Now we can use the Pythagorean Theorem to solve for the hypotenuse. We get: $\sqrt{3^2 + 8^2} = \sqrt{73}$. Now we just have to evaluate $\sin\theta = \dfrac{3}{\sqrt{73}}$. Therefore, $\sin\left(\tan^{-1}\left(\dfrac{3}{8}\right)\right) = \dfrac{3}{\sqrt{73}}$.

Solution to practice problem 9:

Find the value of $\tan\left(\arccos\left(-\frac{6}{7}\right)\right)$.

This problem tells us that we have an angle whose cosine is $-\frac{6}{7}$ and we need to find the tangent of that angle. Let's draw a picture.

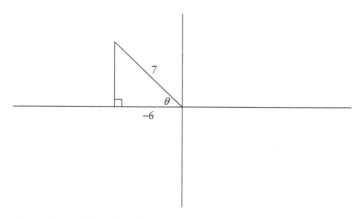

Notice that we located the triangle in Quadrant II. This is because the inverse cosine of a negative number is an angle in Quadrant II. Remember that the tangent of an angle in Quadrant II will be negative. Now we can use the Pythagorean Theorem to solve for the unknown leg. We get: $\sqrt{7^2 - 6^2} = \sqrt{13}$. Now we just have to evaluate $\tan\theta = -\frac{\sqrt{13}}{6}$. Therefore, $\tan\left(\arccos\left(-\frac{6}{7}\right)\right) = -\frac{\sqrt{13}}{6}$.

Solution to practice problem 10:

Find the value of $\cos^{-1}\left(\cos\left(\frac{7\pi}{4}\right)\right)$.

First, let's evaluate $\cos\frac{7\pi}{4} = \frac{\sqrt{2}}{2}$. Now, we just have to evaluate $\cos^{-1}\frac{\sqrt{2}}{2}$. Remember that the inverse cosine of a positive number is an angle in Quadrant I. We know that $\cos\frac{\pi}{4} = \frac{\sqrt{2}}{2}$, so $\cos^{-1}\frac{\sqrt{2}}{2} = \frac{\pi}{4}$. Therefore, $\cos^{-1}\left(\cos\left(\frac{7\pi}{4}\right)\right) = \frac{\pi}{4}$. Notice that the cosine and the inverse did not simply cancel each other and leave you with $\frac{7\pi}{4}$.

UNIT NINE: Inverse Trigonometric Functions

Solution to practice problem 11:
Find the value of $\arcsin\left(\sin\left(-\dfrac{4\pi}{3}\right)\right)$.

First, let's evaluate $\sin\left(-\dfrac{4\pi}{3}\right)$. It might help to draw a picture:

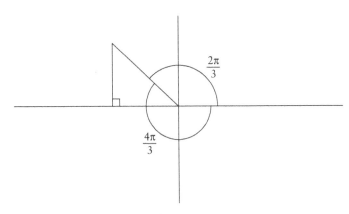

We can see that the negative angle $-\dfrac{4\pi}{3}$ is the same angle as the positive angle $\dfrac{2\pi}{3}$. Here, we get a reference angle of $\dfrac{\pi}{3}$ in Quadrant II, where sine is positive. We get $\sin\left(-\dfrac{4\pi}{3}\right) = \dfrac{\sqrt{3}}{2}$. Next, we have to evaluate $\arcsin\dfrac{\sqrt{3}}{2}$. We know that $\sin\dfrac{\pi}{3} = \dfrac{\sqrt{3}}{2}$, so $\arcsin\dfrac{\sqrt{3}}{2} = \dfrac{\pi}{3}$. Therefore, $\arcsin\left(\sin\left(-\dfrac{4\pi}{3}\right)\right) = \dfrac{\pi}{3}$.

Solution to practice problem 12:
Find the value of $\sin\left(\tan^{-1}\dfrac{x}{2}\right)$, *where x is positive.*

We are given $\tan^{-1}\dfrac{x}{2}$, which means that we can draw a triangle with an angle θ, where $\tan\theta = \dfrac{x}{2}$:

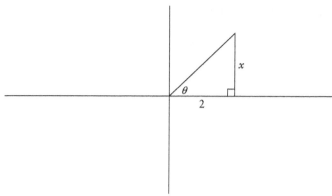

Now we can use the Pythagorean Theorem to solve for the hypotenuse. We get: $\sqrt{x^2+2^2}=\sqrt{x^2+4}$. Now all that we have to do is find $\sin\theta = \dfrac{x}{\sqrt{x^2+4}}$. Therefore, $\sin\left(\tan^{-1}\dfrac{x}{2}\right)=\dfrac{x}{\sqrt{x^2+4}}$.

UNIT TEN
Basic Trigonometric Identities and Equations

Up until now, we have been looking at trigonometric functions as ratios, giving us numerical answers for use in computations. Now we are going to explore the relationships among the six different functions, and what they reveal about trigonometry. These relationships will be very valuable in Calculus and beyond.

Let's go back and look at the definition of sine, cosine, and tangent. As you should recall, given one of the acute angles in a right triangle, the sine of that angle is the $\frac{opposite}{hypotenuse}$, the cosine is the $\frac{adjacent}{hypotenuse}$, and the tangent is the $\frac{opposite}{adjacent}$.

Look at the triangle below:

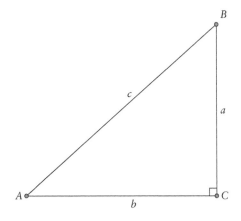

We have $\sin A = \frac{a}{c}$, $\cos A = \frac{b}{c}$, and $\tan A = \frac{a}{b}$. If we take the quotient of sine and cosine, we get $\frac{\sin A}{\cos A} = \frac{\frac{a}{c}}{\frac{b}{c}} = \frac{a}{b} = \tan A$. That is, $\tan A = \frac{\sin A}{\cos A}$. Also, $\cot A = \frac{\cos A}{\sin A}$ because $\cot A = \frac{1}{\tan A}$.

We have the first set of identities from the reciprocal relationships and the quotients:

Reciprocal Identities:

$\sin\theta = \dfrac{1}{\csc\theta}$	$\csc\theta = \dfrac{1}{\sin\theta}$
$\cos\theta = \dfrac{1}{\sec\theta}$	$\sec\theta = \dfrac{1}{\cos\theta}$
$\tan\theta = \dfrac{1}{\cot\theta}$	$\cot\theta = \dfrac{1}{\tan\theta}$

Quotient Identities:

$\tan\theta = \dfrac{\sin\theta}{\cos\theta}$	$\cot\theta = \dfrac{\cos\theta}{\sin\theta}$

We can find some more relationships among these functions. If we apply the Pythagorean Theorem to the triangle in the figure, we get: $a^2 + b^2 = c^2$.

Divide all of the terms by c^2 gives: $\dfrac{a^2}{c^2} + \dfrac{b^2}{c^2} = \dfrac{c^2}{c^2}$.

We can rewrite the expression as: $\left(\dfrac{a}{c}\right)^2 + \left(\dfrac{b}{c}\right)^2 = 1$. Remember that $\sin A = \dfrac{a}{c}$ and $\cos A = \dfrac{b}{c}$. We can substitute to get: $(\sin A)^2 + (\cos A)^2 = 1$.

By the way, it is traditional to write $\sin^2 A$ to mean $(\sin A)^2$ so that there is no confusion as to whether we are squaring the sine function or squaring the angle. Therefore, we can rewrite the identity as: $\sin^2 A + \cos^2 A = 1$. This is the Pythagorean Theorem of Trigonometry. This is a very important relationship, so make sure that you understand and memorize it!

Now let's take this identity and divide all of the terms by $\cos^2 A$.

We get: $\dfrac{\sin^2 A}{\cos^2 A} + \dfrac{\cos^2 A}{\cos^2 A} = \dfrac{1}{\cos^2 A}$. Remember that $\tan A = \dfrac{\sin A}{\cos A}$ and that $\sec A = \dfrac{1}{\cos A}$. This means that we can rewrite this identity as: $\tan^2 A + 1 = \sec^2 A$.

Similarly, we can divide all of the terms by $\sin^2 A$.

We get: $\dfrac{\sin^2 A}{\sin^2 A} + \dfrac{\cos^2 A}{\sin^2 A} = \dfrac{1}{\sin^2 A}$. Because $\cot A = \dfrac{\cos A}{\sin A}$ and $\csc A = \dfrac{1}{\sin A}$, we can rewrite this identity as: $1 + \cot^2 A = \csc^2 A$.

UNIT TEN: Basic Trigonometric Identities and Equations

This gives us another set of identities to memorize:

The Pythagorean Identities:

$\sin^2 \theta + \cos^2 \theta = 1$
$\tan^2 \theta + 1 = \sec^2 \theta$
$1 + \cot^2 \theta = \csc^2 \theta$

Earlier, we showed you the relationships among the cofunctions. Let's show them to you again here so that you have all of the identities all in one place:

$\sin(90° - \theta) = \cos\theta$	$\cos(90° - \theta) = \sin\theta$
$\tan(90° - \theta) = \cot\theta$	$\cot(90° - \theta) = \tan\theta$
$\sec(90° - \theta) = \csc\theta$	$\csc(90° - \theta) = \sec\theta$

Another set of relationships involves odd and even functions. In case you aren't familiar with the concept, an *even* function is one where $f(\theta) = f(-\theta)$. For example, $f(x) = |x|$ is an even function because if you input a positive number or its corresponding negative number for x, you get the same result. For example, $|5| = 5$ and $|-5| = 5$, $|3| = 3$ and $|-3| = 3$, and so on. If you think geometrically, an even function is symmetric across the y-axis. Among the trig functions, cosine and secant are even functions.

An *odd* function is one where $f(\theta) = -f(-\theta)$. For example, $f(x) = x^3$ is an odd function because if you put in a negative number, you get the negative of the result of what you get when you plug in its corresponding positive number. For example, $5^3 = 125$ and $(-5)^3 = -125$, $3^3 = 27$ and $(-3)^3 = -27$, and so on. If you think geometrically, an odd functions is symmetric about the origin. Among the trig functions, sine, cosecant, tangent, and cotangent are odd functions.

Let's put the odd and even functions in a table:

Even/Odd Identities:

Odd	Even
$\sin(-\theta) = -\sin(\theta)$	$\cos(-\theta) = \cos(\theta)$
$\csc(-\theta) = -\csc(\theta)$	$\sec(\theta) = \sec(-\theta)$
$\tan(-\theta) = -\tan(\theta)$	
$\cot(-\theta) = -\cot(\theta)$	

Now that we have learned all of these relationships, let's put them to work. One important use of these trig identities is to simplify complex trig expressions.

Example 1: Let's simplify the following: $\tan x \cos x + \cot x \sin x$.

We know that $\tan x = \dfrac{\sin x}{\cos x}$ and that $\cot x = \dfrac{\cos x}{\sin x}$, so if we substitute these into the expression, we get: $\dfrac{\sin x}{\cos x} \cos x + \dfrac{\cos x}{\sin x} \sin x$.

Now, we can cancel like terms and get: $\sin x + \cos x$.

Therefore, $\tan x \cos x + \cot x \sin x = \sin x + \cos x$.

Let's do another example.

Example 2: Simplify the following: $\dfrac{1-\sin^2 \theta}{\cos \theta}$.

Remember that $\sin^2 \theta + \cos^2 \theta = 1$, which we can rewrite to $\cos^2 \theta = 1 - \sin^2 \theta$.

We can substitute this into our expression to get: $\dfrac{\cos^2 \theta}{\cos \theta} = \cos \theta$.

Let's do a slightly harder one.

Example 3: Simplify the following: $\dfrac{\cot(90° - \theta)\sec \theta}{\sec^2 \theta}$.

Remember that $\cot(90° - \theta) = \tan \theta$. Substitute that into the expression to get: $\dfrac{\tan \theta \sec \theta}{\sec^2 \theta}$.

Next, let's rewrite each of the functions in terms of sine and cosine. We get: $\dfrac{\dfrac{\sin \theta}{\cos \theta} \dfrac{1}{\cos \theta}}{\dfrac{1}{\cos^2 \theta}}$.

Next, we can combine the terms of the fraction in the top: $\dfrac{\dfrac{\sin \theta}{\cos^2 \theta}}{\dfrac{1}{\cos^2 \theta}}$. Finally, we cancel the denominators to get: $\sin \theta$.

Let's do one more.

Example 4: Simplify the following: $\dfrac{\sin x}{1-\cos x} + \dfrac{1-\cos x}{\sin x}$.

How would we combine these in Algebra class? First, we would get a common denominator. Well, we do the same thing here! The common denominator is $\sin x(1-\cos x)$, so we need to multiply the top and bottom of the first expression by $\sin x$ and the top and bottom of the second expression by $(1-\cos x)$. We get: $\dfrac{\sin x}{1-\cos x} \dfrac{\sin x}{\sin x} + \dfrac{1-\cos x}{\sin x} \dfrac{1-\cos x}{1-\cos x}$.

Now let's combine these into one fraction: $\dfrac{\sin^2 x + (1-\cos x)(1-\cos x)}{(1-\cos x)\sin x}$.

Next, let's multiply out $(1-\cos x)(1-\cos x)$: $\dfrac{\sin^2 x + 1 - 2\cos x + \cos^2 x}{(1-\cos x)\sin x}$.

UNIT TEN: Basic Trigonometric Identities and Equations

Remember that $\sin^2 x + \cos^2 x = 1$? Now we can simplify this to: $\dfrac{2-2\cos x}{(1-\cos x)\sin x}$.

Factor out the 2 from the top and cancel: $\dfrac{2(1-\cos x)}{(1-\cos x)\sin x} = \dfrac{2}{\sin x}$. We could rewrite the final answer as $2\csc x$, but that isn't necessary.

Another use of these identities is to solve trigonometric equations.

Example 5: Let's solve for x when $2\sin x \cos x + \sin x = 0$ on the interval $0 \le x < 2\pi$.

If we factor out $\sin x$, we get: $\sin x(2\cos x + 1) = 0$. If you remember from Algebra, this means that either $\sin x = 0$ or $2\cos x + 1 = 0$. The second equation becomes $\cos x = -\dfrac{1}{2}$.

Where does $\sin x = 0$? At $x = 0, \pi$ (on our interval).

Where does $\cos x = -\dfrac{1}{2}$? At $x = \dfrac{2\pi}{3}, \dfrac{4\pi}{3}$.

Let's do another one.

Example 6: Solve for x when $\tan x \sin^2 x = \tan x$ on the interval $0 \le x < 2\pi$.

First, let's subtract $\tan x$ from both sides: $\tan x \sin^2 x - \tan x = 0$.

Next, factor out $\tan x$: $\tan x(\sin^2 x - 1) = 0$. So, either $\tan x = 0$ or $\sin^2 x - 1 = 0$. We can solve the second equation as $\sin x = \pm 1$.

Where does $\tan x = 0$? At $x = 0, \pi$ (on our interval).

Where does $\sin x = \pm 1$? At $x = \dfrac{\pi}{2}, \dfrac{3\pi}{2}$.

Time to practice!

Practice Problems

Practice problem 1: Simplify the following: $\sin x + \sin(-x) + \cos x + \cos(-x)$.

Practice problem 2: Simplify the following: $\sin\theta - \tan\theta\cos\theta + \cos(90° - \theta)$.

Practice problem 3: Simplify the following: $\dfrac{\sec^2 A - \tan^2 A}{\sin^2 A + \cos^2 A}$.

Practice problem 4: Simplify the following: $\dfrac{(\sin x + \cos x)^2 - 1}{\sin x \cos x}$.

Practice problem 5: Simplify the following: $\dfrac{\tan x}{\csc^2 x} + \dfrac{\tan x}{\sec^2 x}$.

Practice problem 6: Simplify the following: $\sin x \cos x \tan x \sec x \csc x$.

Practice problem 7: Simplify the following: $1 - \dfrac{1-\cos^2\theta}{1+\cos\theta}$.

Practice problem 8: Simplify the following: $\sqrt{\cos^2 x + 1 + \dfrac{2}{\sec x}}$.

Practice problem 9: Solve for x when $\tan x - \sqrt{2}\tan x \sin x = 0$ on the interval $0 \le x < 2\pi$.

Practice problem 10: Solve for x when $\sec^2 x - 4 = 0$ on the interval $0 \le x < 2\pi$.

Practice problem 11: Solve for θ when $2\sin^2 \theta + 3\sin \theta + 1 = 0$ on the interval $0° \le \theta < 360°$.

Practice problem 12: Solve for θ when $\cos^2 \theta - 2\cos \theta = 0$ on the interval $0° \le \theta < 360°$.

Solutions to the Practice Problems

Solution to practice problem 1: *Simplify the following:* $\sin x + \sin(-x) + \cos x + \cos(-x)$.

Remember that $\sin(-x) = -\sin x$ and that $\cos(-x) = \cos x$. Now we can substitute into the expression to get: $\sin x + (-\sin x) + \cos x + (\cos x)$. The sines cancel, and we get $2\cos x$.

Solution to practice problem 2: *Simplify the following:* $\sin \theta - \tan \theta \cos \theta + \cos(90° - \theta)$.

Remember that $\tan \theta = \dfrac{\sin \theta}{\cos \theta}$ and that $\cos(90° - \theta) = \sin \theta$. Now we can substitute into the expression to get: $\sin \theta - \dfrac{\sin \theta}{\cos \theta}\cos \theta + \sin \theta$. Next, we can cancel the cosines to get: $\sin \theta - \sin \theta + \sin \theta$, which simplifies to $\sin \theta$.

Solution to practice problem 3: *Simplify the following:* $\dfrac{\sec^2 A - \tan^2 A}{\sin^2 A + \cos^2 A}$.

Remember the Pythagorean Identities $\sin^2 A + \cos^2 A = 1$ and $1 + \tan^2 A = \sec^2 A$. Now, we can substitute into the expression to get: $\dfrac{1 + \tan^2 A - \tan^2 A}{1}$, which reduces to 1.

Solution to practice problem 4: *Simplify the following:* $\dfrac{(\sin x + \cos x)^2 - 1}{\sin x \cos x}$.

Let's multiply out the expression on top to get: $\dfrac{\sin^2 x + 2\sin x \cos x + \cos^2 x - 1}{\sin x \cos x}$.

Next, use the Pythagorean Identity $\sin^2 x + \cos^2 x = 1$ and substitute into the numerator: $\dfrac{1 + 2\sin x \cos x - 1}{\sin x \cos x}$. Now we can simplify this to: $\dfrac{2\sin x \cos x}{\sin x \cos x}$, which reduces to 2.

UNIT TEN: Basic Trigonometric Identities and Equations

Solution to practice problem 5: *Simplify the following:* $\dfrac{\tan x}{\csc^2 x} + \dfrac{\tan x}{\sec^2 x}$.

First, let's use the identities $\sec x = \dfrac{1}{\cos x}$ and $\csc x = \dfrac{1}{\sin x}$. Now, we can substitute those identities into the expression to get: $\dfrac{\tan x}{\frac{1}{\sin^2 x}} + \dfrac{\tan x}{\frac{1}{\cos^2 x}}$. This can be rewritten as: $\tan x \sin^2 x + \tan x \cos^2 x$. Next, factor out $\tan x$: $\tan x (\sin^2 x + \cos^2 x)$. Next, use the Pythagorean Identity $\sin^2 x + \cos^2 x = 1$ and substitute: $(\tan x)(1) = \tan x$.

Solution to practice problem 6: *Simplify the following:* $\sin x \cos x \tan x \sec x \csc x$.

Remember that $\sec x = \dfrac{1}{\cos x}$ and $\csc x = \dfrac{1}{\sin x}$. Now we can substitute those identities into the expression to get: $\sin x \cos x \tan x \dfrac{1}{\cos x}\dfrac{1}{\sin x}$. Now, we can do some cancelling, and we are left with: $\tan x$.

Solution to practice problem 7: *Simplify the following:* $1 - \dfrac{1-\cos^2 \theta}{1+\cos \theta}$.

First, factor the numerator of the expression: $1 - \dfrac{(1-\cos\theta)(1+\cos\theta)}{1+\cos\theta}$. Now, we can cancel the like terms to get: $1 - (1-\cos\theta)$, which simplifies to $\cos\theta$.

Solution to practice problem 8: *Simplify the following:* $\sqrt{\cos^2 x + 1 + \dfrac{2}{\sec x}}$.

Remember that $\sec x = \dfrac{1}{\cos x}$. We can substitute into the expression to get: $\sqrt{\cos^2 x + 1 + \dfrac{2}{\frac{1}{\cos x}}}$, which we can rewrite as $\sqrt{\cos^2 x + 1 + 2\cos x}$. This can be factored as $\sqrt{(\cos x + 1)^2}$. Now take the square root to get: $\cos x + 1$.

Solution to practice problem 9: *Solve for x when* $\tan x - \sqrt{2}\tan x \sin x = 0$ *on the interval* $0 \leq x < 2\pi$.

If we factor out $\tan x$, we get: $\tan x(1 - \sqrt{2}\sin x) = 0$. This means that either $\tan x = 0$ or $1 - \sqrt{2}\sin x = 0$. The second equation becomes $\sin x = -\dfrac{1}{\sqrt{2}}$. Don't forget that $\dfrac{1}{\sqrt{2}} = \dfrac{\sqrt{2}}{2}$.

Where does $\tan x = 0$? At $x = 0,\ \pi$ (on our interval).

Where does $\sin x = -\frac{1}{\sqrt{2}}$? At $x = \frac{5\pi}{4},\ \frac{7\pi}{4}$.

Solution to practice problem 10: *Solve for x when $\sec^2 x - 4 = 0$ on the interval $0 \le x < 2\pi$.*

First, let's factor $\sec^2 x - 4 = 0$ into: $(\sec x + 2)(\sec x - 2) = 0$.

So, either $\sec x = 2$ or $\sec x = -2$, which means that $\cos x = \frac{1}{2}$ or $\cos x = -\frac{1}{2}$.

Where does $\cos x = \frac{1}{2}$? At $x = \frac{\pi}{3},\ \frac{5\pi}{3}$ (on our interval).

Where does $\cos x = -\frac{1}{2}$? At $x = \frac{2\pi}{3},\ \frac{4\pi}{3}$.

Solution to practice problem 11: *Solve for θ when $2\sin^2 \theta + 3\sin \theta + 1 = 0$ on the interval $0° \le \theta < 360°$.*

First, let's factor $2\sin^2 \theta + 3\sin \theta + 1 = 0$ into: $(2\sin \theta + 1)(\sin \theta + 1) = 0$.

So, either $2\sin \theta + 1 = 0$ or $\sin \theta + 1 = 0$ which means that $\sin \theta = -\frac{1}{2}$ or $\sin \theta = -1$.

Where does $\sin \theta = -\frac{1}{2}$? At $\theta = 210°,\ 330°$ (on our interval).

Where does $\sin \theta = -1$? At $\theta = 270°$.

Solution to practice problem 12: *Solve for θ when $\cos^2 \theta - 2\cos \theta = 0$ on the interval $0° \le \theta < 360°$.*

First, let's factor $\cos^2 \theta - 2\cos \theta = 0$ into: $\cos \theta (\cos \theta - 2) = 0$.

So, either $\cos \theta = 0$ or $\cos \theta = 2$.

Where does $\cos \theta = 0$? At $\theta = 90°,\ 270°$ (on our interval).

Where does $\cos \theta = 2$? Nowhere! Remember that cosine is always between -1 and 1. So there is no solution for the second equation.

UNIT ELEVEN
More Trigonometric Identities

Now we are going to explore an area of Trigonometry that you will either love or hate. We are going to use the various trig relationships that we have learned to do proofs. These are similar to the simplifications that we did in the previous unit except that this time we will be given a trigonometric expression that we are told is equal to another trigonometric expression. Our job will be to show that the expression on the left side of the equals sign is the same as the one on the right side. These are little puzzles that will call on us to remember the relationships and to use Algebra to get the two expressions to equal each other. We can manipulate the expression on the left or on the right. What we are NOT allowed to do is to move an expression from one side to the other.

Let's do an example.

Example 1: Verify the following identity: $(\cos x)(\tan x + \sin x \cot x) = \sin x + \cos^2 x$.

There isn't much that we can do with the right side, so let's concentrate on the left side.

First, let's distribute $\cos x$: $\cos x \tan x + \cos x \sin x \cot x$.

Next, let's use the relationships $\tan x = \dfrac{\sin x}{\cos x}$ and $\cot x = \dfrac{\cos x}{\sin x}$ to rewrite the left side: $\cos x \dfrac{\sin x}{\cos x} + \cos x \sin x \dfrac{\cos x}{\sin x}$.

Now we simplify: $\sin x + \cos x \cos x$

Which we can rewrite as: $\sin x + \cos^2 x$.

And there we go. We have proved our first identity!

Let's do another one.

Example 2: Verify the following identity: $\dfrac{\sin^2 x + \cos^2 x}{\csc x} = \sin x$.

First, remember that $\sin^2 x + \cos^2 x = 1$, so we can rewrite the left side: $\dfrac{1}{\csc x}$.

Next, let's use the relationship $\csc x = \dfrac{1}{\sin x}$ to rewrite the expression: $\dfrac{1}{\frac{1}{\sin x}}$, which we can now rewrite as $\sin x$.

That wasn't so hard! Here is another one:

Example 3: Verify the following identity: $\sin^2 x - \cos^2 x = 1 - 2\cos^2 x$.

It isn't obvious whether we should start with the left side or the right, so let's start with the left. When you see a trig function that is squared, your first instinct should be to try one of the Pythagorean relationships. Here, we can use the Pythagorean relationship $\sin^2 x + \cos^2 x = 1$, which we can transform into $\sin^2 x = 1 - \cos^2 x$. We get: $(1 - \cos^2 x) - \cos^2 x$, which simplifies to $1 - 2\cos^2 x$.

That was easy! Let's do a slightly harder one.

Example 4: Verify the following identity: $\tan^2 x - \sin^2 x = \tan^2 x \sin^2 x$.

Here we could use the Pythagorean relationships, but all that would happen is that we could be converting one trig function into a different one. We already have the ones that we need on both sides of the identity. So, when in doubt, convert everything that is not a sine or a cosine into a different function that is a sine or a cosine. We can use the relationship $\tan x = \dfrac{\sin x}{\cos x}$ to rewrite both sides: $\dfrac{\sin^2 x}{\cos^2 x} - \sin^2 x = \dfrac{\sin^2 x}{\cos^2 x} \sin^2 x$.

Now what? Let's get a common denominator for the terms on the left side. Let's multiply $\sin^2 x$ by $\dfrac{\cos^2 x}{\cos^2 x}$ to get: $\dfrac{\sin^2 x}{\cos^2 x} - \sin^2 x \dfrac{\cos^2 x}{\cos^2 x} =$.

We can now combine the numerators: $\dfrac{\sin^2 x - \sin^2 x \cos^2 x}{\cos^2 x} =$.

Next, factor $\sin^2 x$ out of the numerator: $\dfrac{\sin^2 x (1 - \cos^2 x)}{\cos^2 x} =$.

Remember that $\sin^2 x = 1 - \cos^2 x$? Now we can rewrite this to get: $\dfrac{\sin^2 x (\sin^2 x)}{\cos^2 x} =$.

We can break this into: $\dfrac{\sin^2 x}{\cos^2 x}(\sin^2 x) =$.

Finally, because $\tan x = \dfrac{\sin x}{\cos x}$, we can rewrite this as: $\tan^2 x \sin^2 x =$.

By the way, did you notice that multiplying $\sin^2 x$ and $\tan^2 x$ is the same as subtracting them? You will find that there are some very interesting relationships among the different trig functions.

How about this one?

Example 5: Verify the following identity: $\dfrac{1 + \cos x}{1 - \cos x} = \dfrac{\sec x + 1}{\sec x - 1}$.

This time, we will work with the right side. Why? Because we are going to want to convert the secants into cosines to see what that does for us. Using the relationship $\sec x = \dfrac{1}{\cos x}$, we get: $= \dfrac{\dfrac{1}{\cos x} + 1}{\dfrac{1}{\cos x} - 1}$.

Next, we can multiply 1 by $\dfrac{\cos x}{\cos x}$ to get a common denominator: $= \dfrac{\dfrac{1}{\cos x} + \dfrac{\cos x}{\cos x}}{\dfrac{1}{\cos x} - \dfrac{\cos x}{\cos x}}$

UNIT ELEVEN: More Trigonometric Identities

Now, combine the terms into single fractions; $= \dfrac{\frac{1+\cos x}{\cos x}}{\frac{1-\cos x}{\cos x}}$. Now we just cancel the denominators $= \dfrac{1+\cos x}{1-\cos x}$, and we are done.

Now let's do a more difficult one:

Example 6: Verify the following identity: $\tan x + \sec x = \dfrac{\frac{1}{\sec x}}{1-\sin x}$.

Let's start with the left side first. Let's use the relationships $\tan x = \dfrac{\sin x}{\cos x}$ and $\sec x = \dfrac{1}{\cos x}$ to rewrite the left side: $\dfrac{\sin x}{\cos x} + \dfrac{1}{\cos x} =$. These have a common denominator of $\cos x$ so let's combine the numerators: $\dfrac{1+\sin x}{\cos x} =$. We seem to have run out of things to do on the left side, so let's work on the right side. We can again use the relationship $\sec x = \dfrac{1}{\cos x}$ to rewrite the right side: $= \dfrac{\frac{1}{\frac{1}{\cos x}}}{1-\sin x}$, which we can rewrite to: $= \dfrac{\cos x}{1-\sin x}$.

Now what do we do? Notice that if we were allowed to cross-multiply, we would get: $(1-\sin x)(1+\sin x) = \cos^2 x$, which we could then simplify to $1-\sin^2 x = \cos^2 x$, which we know is true. Unfortunately, we are NOT allowed to cross-multiply. But, now that we have seen this relationship, we have a strategy to do something similar.

Here is what we can do. Multiply the expression on the left side by $\dfrac{1-\sin x}{1-\sin x}$. Now we get: $\dfrac{1+\sin x}{\cos x} \dfrac{1-\sin x}{1-\sin x} =$.

We can combine these into: $\dfrac{1-\sin^2 x}{(\cos x)(1-\sin x)} =$.

Now, we can use the relationship $1-\sin^2 x = \cos^2 x$ and substitute into the numerator: $\dfrac{\cos^2 x}{(\cos x)(1-\sin x)} =$.

Now we cancel a $\cos x$ term: $\dfrac{\cos x}{1-\sin x} =$.

And we have proved the identity! Notice some of the things that we did. We worked with both sides of the identity. We used a common denominator to rewrite an expression into one that we could simplify. And we did not distribute the terms in the denominator of the expression. Why not? Because we want to first see if we can cancel something. Beware of taking two uncomplicated expressions and turning them into a single, more complicated one!

Now it's your turn!

Practice Problems

Verify the following identities.

Practice problem 1: $\dfrac{\tan x}{\sec x} = \sin x$

Practice problem 2: $\dfrac{1-\cos^2 x}{\cos x} = \tan x \sin x$

Practice problem 3: $\dfrac{1}{1-\cos x} + \dfrac{1}{1+\cos x} = 2\csc^2 x$

Practice problem 4: $\tan x + \dfrac{1}{\tan x} = \sec x \csc x$

Practice problem 5: $\dfrac{\sec x + 1}{\tan x} = \dfrac{\sin x}{1-\cos x}$

Practice problem 6: $\tan^4 x + \tan^2 x = \sec^4 x - \sec^2 x$

Practice problem 7: $\dfrac{\sin x + \cos x}{\sin x - \cos x} = \dfrac{1+2\sin x \cos x}{2\sin^2 x - 1}$

Practice problem 8: $\sin^3 x \cos^3 x = (\sin^3 x - \sin^5 x)(\cos x)$

Practice problem 9: $\dfrac{\cos x + \sin x}{\cos x - \sin x} = \dfrac{2\tan x}{1-\tan^2 x} + \dfrac{1}{2\cos^2 x - 1}$

Practice problem 10: $(A\sin x + B\cos x)^2 + (A\cos x - B\sin x)^2 = A^2 + B^2$

Practice problem 11: $\dfrac{\sin A \cos B + \cos A \sin B}{\cos A \cos B - \sin A \sin B} = \dfrac{\tan A + \tan B}{1 - \tan A \tan B}$

Practice problem 12: $(\sin^2 x - \cos^2 x)(1 + \tan^2 x) = \tan^2 x - 1$

UNIT ELEVEN: More Trigonometric Identities

Solutions to the Practice Problems

Verify the following identities:

Solution to practice problem 1: $\dfrac{\tan x}{\sec x} = \sin x$

Here, there is nothing to do with the right side, so let's work on the left. We can use the relationships $\tan x = \dfrac{\sin x}{\cos x}$ and $\sec x = \dfrac{1}{\cos x}$ to get: $\dfrac{\frac{\sin x}{\cos x}}{\frac{1}{\cos x}} =$. Now we simply cancel the denominators: $\dfrac{\sin x}{1} =$, and we're done.

Solution to practice problem 2: $\dfrac{1-\cos^2 x}{\cos x} = \tan x \sin x$

First, let's use the relationship $\tan x = \dfrac{\sin x}{\cos x}$ to rewrite the right side as: $= \dfrac{\sin x}{\cos x} \sin x$, which simplifies to $= \dfrac{\sin^2 x}{\cos x}$. Next, remember that $\sin^2 x = 1 - \cos^2 x$ to rewrite this as: $= \dfrac{1-\cos^2 x}{\cos x}$.

Solution to practice problem 3: $\dfrac{1}{1-\cos x} + \dfrac{1}{1+\cos x} = 2\csc^2 x$

Let's combine the two fractions using a common denominator. We multiply the left one by $\dfrac{1+\cos x}{1+\cos x}$ and the right one by $\dfrac{1-\cos x}{1-\cos x}$. We get: $\dfrac{1}{1-\cos x}\dfrac{1+\cos x}{1+\cos x} + \dfrac{1}{1+\cos x}\dfrac{1-\cos x}{1-\cos x} =$. Now, let's simplify the denominator and combine the fractions: $\dfrac{1+\cos x+1-\cos x}{1-\cos^2 x} =$.
This simplifies to: $\dfrac{2}{1-\cos^2 x} =$. Next, remember that $\sin^2 x = 1-\cos^2 x$, so we can rewrite this as $\dfrac{2}{\sin^2 x} =$. Finally, we know that $\csc x = \dfrac{1}{\sin x}$, so we can rewrite this as $2\csc^2 x =$.

Solution to practice problem 4: $\tan x + \dfrac{1}{\tan x} = \sec x \csc x$

Let's work with the left side first. We know that $\cot x = \dfrac{1}{\tan x}$, so we get: $\tan x + \cot x =$. Next, we can use the relationships $\tan x = \dfrac{\sin x}{\cos x}$ and $\cot x = \dfrac{\cos x}{\sin x}$

to rewrite this as $\frac{\sin x}{\cos x} + \frac{\cos x}{\sin x} =$. Now what? Let's combine the two fractions with a common denominator. We will multiply the one on the left by $\frac{\sin x}{\sin x}$ and the one on the right by $\frac{\cos x}{\cos x}$. We get: $\frac{\sin x \sin x}{\cos x \sin x} + \frac{\cos x \cos x}{\sin x \cos x} =$. Now we can combine the two fractions: $\frac{\sin^2 x + \cos^2 x}{\sin x \cos x} =$. Remember that $\sin^2 x + \cos^2 x = 1$! This gives us $\frac{1}{\sin x \cos x} =$. Finally, we can use the relationships $\sec x = \frac{1}{\cos x}$ and $\csc x = \frac{1}{\sin x}$ to get $\sec x \csc x =$.

Solution to practice problem 5: $\frac{\sec x + 1}{\tan x} = \frac{\sin x}{1 - \cos x}$

Let's work with the left side first. We know that $\tan x = \frac{\sin x}{\cos x}$ and $\sec x = \frac{1}{\cos x}$, so we can rewrite the left side to $\frac{\frac{1}{\cos x} + 1}{\frac{\sin x}{\cos x}} =$. Next, let's multiply 1 by $\frac{\cos x}{\cos x}$: $\frac{\frac{1}{\cos x} + \frac{\cos x}{\cos x}}{\frac{\sin x}{\cos x}} =$. Now we can combine the fractions in the numerator: $\frac{\frac{1 + \cos x}{\cos x}}{\frac{\sin x}{\cos x}} =$. Then we cancel the denominators to get: $\frac{1 + \cos x}{\sin x} =$. We have seen something similar to this in the last example before the practice problems. Multiply the expression on the left side by $\frac{1 - \cos x}{1 - \cos x}$. Now we get: $\frac{1 + \cos x}{\sin x} \times \frac{1 - \cos x}{1 - \cos x} =$.

We can combine these into: $\frac{1 - \cos^2 x}{(\sin x)(1 - \cos x)} =$.

Now, we can use the relationship $1 - \cos^2 x = \sin^2 x$ and substitute into the numerator: $\frac{\sin^2 x}{(\sin x)(1 - \cos x)} =$.

Now we cancel a $\sin x$ term: $\frac{\sin x}{1 - \cos x} =$.

Solution to practice problem 6: $\tan^4 x + \tan^2 x = \sec^4 x - \sec^2 x$

Let's work with the right side first. Let's factor out $\sec^2 x$: $= \sec^2 x (\sec^2 x - 1)$. Next, let's use the Pythagorean identity $1 + \tan^2 x = \sec^2 x$ and substitute: $= (1 + \tan^2 x)((1 + \tan^2 x) - 1)$.

Now we simplify: $= (1 + \tan^2 x)(\tan^2 x)$. Then we distribute: $= \tan^2 x + \tan^4 x$.

UNIT ELEVEN: More Trigonometric Identities 127

Solution to practice problem 7: $\dfrac{\sin x + \cos x}{\sin x - \cos x} = \dfrac{1 + 2\sin x \cos x}{2\sin^2 x - 1}$

Let's work with the left side first. When you see the sum or difference of two trig functions, or a number and a trig function, try multiplying the numerator and the denominator by the conjugate of the denominator. This will get you squared terms, which means that we can then use Pythagorean identities to substitute. Here, we will multiply by $\dfrac{\sin x + \cos x}{\sin x + \cos x}$. We get: $\dfrac{\sin x + \cos x}{\sin x - \cos x} \dfrac{\sin x + \cos x}{\sin x + \cos x} =$. Now, distribute the terms in the numerator and the denominator: $\dfrac{\sin^2 x + 2\sin x \cos x + \cos^2 x}{\sin^2 x - \cos^2 x} =$.

Now we can use the Pythagorean identity $\sin^2 x + \cos^2 x = 1$ in the numerator: $\dfrac{1 + 2\sin x \cos x}{\sin^2 x - \cos^2 x} =$. Use another Pythagorean identity $\cos^2 x = 1 - \sin^2 x$ in the denominator: $\dfrac{1 + 2\sin x \cos x}{\sin^2 x - (1 - \sin^2 x)} =$. This simplifies to $\dfrac{1 + 2\sin x \cos x}{2\sin^2 x - 1} =$.

Solution to practice problem 8: $\sin^3 x \cos^3 x = (\sin^3 x - \sin^5 x)(\cos x)$

Let's work with the terms on the right. First, let's factor out $\sin^3 x := \sin^3 x (1 - \sin^2 x)(\cos x)$. Next, let's use the Pythagorean relationship $\sin^2 x = 1 - \cos^2 x$ and substitute: $= \sin^3 x (1 - (1 - \cos^2 x))(\cos x)$, which simplifies to $= \sin^3 x (\cos^2 x)(\cos x)$. And this simplifies to $= \sin^3 x \cos^3 x$.

Solution to practice problem 9: $\dfrac{\cos x + \sin x}{\cos x - \sin x} = \dfrac{2\tan x}{1 - \tan^2 x} + \dfrac{1}{2\cos^2 x - 1}$

Let's work with the terms on the left. Here, we will multiply by the conjugate of the denominator, $\dfrac{\cos x + \sin x}{\cos x + \sin x}$, to get: $\dfrac{\cos x + \sin x}{\cos x - \sin x} \dfrac{\cos x + \sin x}{\cos x + \sin x} =$.

Distribute: $\dfrac{\cos^2 x + 2\sin x \cos x + \sin^2 x}{\cos^2 x - \sin^2 x} =$. Next use the Pythagorean Identity $\sin^2 x + \cos^2 x = 1$ to simplify the numerator: $\dfrac{2\sin x \cos x + 1}{\cos^2 x - \sin^2 x} =$.

Now we can break this into two fractions: $\dfrac{2\sin x \cos x}{\cos^2 x - \sin^2 x} + \dfrac{1}{\cos^2 x - \sin^2 x} =$.

Let's use the Pythagorean relationship $\sin^2 x = 1 - \cos^2 x$ on the right-hand term to get: $\dfrac{2\sin x \cos x}{\cos^2 x - \sin^2 x} + \dfrac{1}{\cos^2 x - (1 - \cos^2 x)} =$, which simplifies to $\dfrac{2\sin x \cos x}{\cos^2 x - \sin^2 x} + \dfrac{1}{2\cos^2 x - 1} =$. Almost done! Now here is a neat trick! In the left-hand term, if we divide the terms in the numerator and denominator by $\cos^2 x$, we

get: $\dfrac{\dfrac{\cos^2 x}{\cos^2 x}}{\dfrac{\cos^2 x}{\cos^2 x} - \dfrac{\sin^2 x}{\cos^2 x}} + \dfrac{1}{2\cos^2 x - 1} =$, which simplifies to $\dfrac{\dfrac{2\sin x}{\cos x}}{1 - \dfrac{\sin^2 x}{\cos^2 x}} + \dfrac{1}{2\cos^2 x - 1} =$.

Finally, we can use the relationship $\tan x = \dfrac{\sin x}{\cos x}$ and get: $\dfrac{2\tan x}{1 - \tan^2 x} + \dfrac{1}{2\cos^2 x - 1} =$.

Solution to practice problem 10: $(A\sin x + B\cos x)^2 + (A\cos x - B\sin x)^2 = A^2 + B^2$

First, let's multiply out the left side: $A^2 \sin^2 x + 2AB\sin x \cos x + B^2 \cos^2 x + A^2 \cos^2 x - 2AB\sin x \cos x + B^2 \sin^2 x =$, which simplifies to: $A^2 \sin^2 x + B^2 \cos^2 x + A^2 \cos^2 x + B^2 \sin^2 x =$. We can rearrange this into: $A^2 \sin^2 x + A^2 \cos^2 x + B^2 \cos^2 x + B^2 \sin^2 x =$. Next, factor out A^2 and B^2: $A^2 \left(\sin^2 x + \cos^2 x \right) + B^2 \left(\sin^2 x + \cos^2 x \right) =$. Finally, use the Pythagorean relationship $\sin^2 x + \cos^2 x = 1$: $A^2 + B^2 =$.

Solution to practice problem 11: $\dfrac{\sin A \cos B + \cos A \sin B}{\cos A \cos B - \sin A \sin B} = \dfrac{\tan A + \tan B}{1 - \tan A \tan B}$

This one requires a trick that you probably won't think of. We are going to divide each of the four terms on the left side by $\cos A \cos B$. This gives us: $\dfrac{\dfrac{\sin A \cos B}{\cos A \cos B} + \dfrac{\cos A \sin B}{\cos A \cos B}}{\dfrac{\cos A \cos B}{\cos A \cos B} - \dfrac{\sin A \sin B}{\cos A \cos B}} =$. Which simplifies to: $\dfrac{\dfrac{\sin A}{\cos A} + \dfrac{\sin B}{\cos B}}{1 - \dfrac{\sin A \sin B}{\cos A \cos B}} =$. Finally, use the relationship $\tan x = \dfrac{\sin x}{\cos x}$ to get: $\dfrac{\tan A + \tan B}{1 - \tan A \tan B} =$.

Solution to practice problem 12: $\left(\sin^2 x - \cos^2 x \right)\left(1 + \tan^2 x \right) = \tan^2 x - 1$

First, let's use the Pythagorean identity $1 + \tan^2 x = \sec^2 x$ and substitute into the left side: $\left(\sin^2 x - \cos^2 x \right)\left(\sec^2 x \right) =$. Next, let's use the relationship $\sec x = \dfrac{1}{\cos x}$ and substitute: $(\sin^2 x - \cos^2 x)\left(\dfrac{1}{\cos^2 x} \right) =$. Distribute: $\dfrac{\sin^2 x}{\cos^2 x} - \dfrac{\cos^2 x}{\cos^2 x} =$. Finally, use the relationship $\tan x = \dfrac{\sin x}{\cos x}$ to simplify: $\tan^2 x - 1 =$.

UNIT TWELVE
Trigonometric Angle Formulas

Now we are going to learn some formulas that work with the trig functions of more than one angle, double angles, or half angles. These will enable us to do some interesting calculations and will show some new relationships among the different trig functions.

Angle Sum and Difference Formulas

Suppose we want to find the $\sin(A+B)$, where A and B are two angles that we know. We might be tempted to "distribute the sine" to get $\sin(A+B) = \sin A + \sin B$, but this is incorrect. We can see that this is not true with an easy example: $\sin(30° + 30°) = \sin 60° = \frac{\sqrt{3}}{2}$, but $\sin 30° + \sin 30° = \frac{1}{2} + \frac{1}{2} = 1$. It is a common mistake to think of sine this way but remember that we are not finding the "sine *times* A plus B." We are finding the "sine *of* A and B." This is something very different.

So what is $\sin(A+B)$? It turns out that $\sin(A+B) = \sin A \cos B + \cos A \sin B$. We are not going to derive this formula; just take our word for it!

Let's do an example.

Example 1: Suppose we want to evaluate $\sin 75°$. How would we do it?

We know that $75° = 45° + 30$, and we know the trig functions of both $45°$ and $30°$. This means that we can use our new formula: $\sin(A+B) = \sin A \cos B + \cos A \sin B$. We get: $\sin 75° = \sin(45° + 30°) = \sin 45° \cos 30° + \cos 45° \sin 30°$. Now, we can plug in the trig values: $\sin(45° + 30°) = \frac{\sqrt{2}}{2}\frac{\sqrt{3}}{2} + \frac{\sqrt{2}}{2}\frac{1}{2} = \frac{\sqrt{6} + \sqrt{2}}{4}$. Check it on your calculator!

If we take our formula and we replace B with $-B$, we can find $\sin(A-B)$. We get: $\sin(A-B) = \sin A \cos(-B) + \cos A \sin(-B)$. We also know that $\cos(-B) = \cos B$ and $\sin(-B) = -\sin B$. If we substitute these into our formula, we get: $\sin(A-B) = \sin A \cos B + \cos A(-\sin B)$, which can be simplified to: $\sin(A-B) = \sin A \cos B - \cos A \sin B$.

Let's do an example.

Example 2: Suppose we want to evaluate $\sin 15°$. How would we do it?

We know that $15° = 45° - 30°$ and we know the trig functions of both $45°$ and $30°$. This means that we can use our new formula: $\sin(A-B) = \sin A \cos B - \cos A \sin B$.

We get: $\sin 15° = \sin(45° - 30°) = \sin 45° \cos 30° - \cos 45° \sin 30°$. Now we can plug in the trig values: $\sin(45° - 30°) = \dfrac{\sqrt{2}}{2}\dfrac{\sqrt{3}}{2} - \dfrac{\sqrt{2}}{2}\dfrac{1}{2} = \dfrac{\sqrt{6}-\sqrt{2}}{4}$.

What about if we want to find the cosine of the sum of two angles? Now the formula changes a bit: $\cos(A+B) = \cos A \cos B - \sin A \sin B$.

Let's do an example.

Example 3: Evaluate $\cos 75°$.

As in Example 1, we will find $\cos 75°$ by evaluating $\cos(45° + 30°)$ and using the formula that we just learned. Here we get: $\cos(45° + 30°) = \cos 45° \cos 30° - \sin 45° \sin 30°$. Now we can plug in the trig values and get: $\cos(45° + 30°) = \dfrac{\sqrt{2}}{2}\dfrac{\sqrt{3}}{2} - \dfrac{\sqrt{2}}{2}\dfrac{1}{2} = \dfrac{\sqrt{6}-\sqrt{2}}{4}$. Did you notice that this is the same as $\sin 15°$? This is because $\cos 75° = \sin(90° - 75°) = \sin 15°$.

What about the cosine of the difference of two angles? As with the sine formula, the only thing that changes is the sign in the middle of the formula. Here we get: $\cos(A - B) = \cos A \cos B + \sin A \sin B$.

Example 4: Evaluate $\cos 15°$.

As in Example 2, we will find $\cos 15°$ by evaluating $\cos(45° - 30°)$ and using the formula that we just learned. Here we get: $\cos(45° - 30°) = \cos 45° \cos 30° + \sin 45° \sin 30°$. Now we can plug in the trig values and get: $\cos(45° - 30°) = \dfrac{\sqrt{2}}{2}\dfrac{\sqrt{3}}{2} + \dfrac{\sqrt{2}}{2}\dfrac{1}{2} = \dfrac{\sqrt{6}+\sqrt{2}}{4}$. Did you notice that this is the same as $\sin 75°$? This is because $\cos 15° = \sin(90° - 15°) = \sin 75°$ and $\cos 75° = \sin(90° - 75°) = \sin 15°$.

Now let's learn the Angle Sum and Difference formulas for tangents.

If you want to find the tangent of the sum of two angles, the formula is:

$$\tan(A+B) = \dfrac{\tan A + \tan B}{1 - \tan A \tan B}.$$

Similarly, the formula for the tangent of the difference of two angles is:

$$\tan(A-B) = \dfrac{\tan A - \tan B}{1 + \tan A \tan B}.$$

Example 5: Find $\tan 75°$.

Just as with the sine and cosine of 75°, we use the two angles 45° and 30° in the formula. We get: $\tan 75° = \tan(45° + 30°) = \dfrac{\tan 45° + \tan 30°}{1 - \tan 45° \tan 30°}$. Now we can plug in the trig values and get: $\tan(45° + 30°) = \dfrac{1 + \dfrac{1}{\sqrt{3}}}{1 - (1)\left(\dfrac{1}{\sqrt{3}}\right)}$, which, with a little algebra, simplifies to $\dfrac{\sqrt{3} + 1}{\sqrt{3} - 1}$.

To recap, the six formulas that we just learned are:

Angle Sum and Difference Formulas:

$$\sin(A + B) = \sin A \cos B + \cos A \sin B$$

$$\sin(A - B) = \sin A \cos B - \cos A \sin B$$

$$\cos(A + B) = \cos A \cos B - \sin A \sin B$$

$$\cos(A - B) = \cos A \cos B + \sin A \sin B$$

$$\tan(A + B) = \frac{\tan A + \tan B}{1 - \tan A \tan B}$$

$$\tan(A - B) = \frac{\tan A - \tan B}{1 + \tan A \tan B}$$

Double Angle Formulas

Now we are going to learn what are called the "Double Angle Formulas."

Suppose we know $\sin A$ and we want to find $\sin 2A$. As you can probably guess by now, $\sin 2A$ is not the same as $2 \sin A$. Instead, we can find this with the angle addition formula. We start with $\sin(A + B) = \sin A \cos B + \cos A \sin B$ and now let $B = A$. This gives us: $\sin(A + A) = \sin A \cos A + \cos A \sin A$, which can be simplified to $\sin 2A = 2 \sin A \cos A$.

Now let's find $\cos 2A$ using the same approach. We start with: $\cos(A + B) = \cos A \cos B - \sin A \sin B$ and now let $B = A$. This gives us: $\cos(A + A) = \cos A \cos A - \sin A \sin A$, which can be simplified to $\cos 2A = \cos^2 A - \sin^2 A$.

Let's do a couple of examples.

Example 6: If $\sin A = \dfrac{3}{5}$ and angle A is in Quadrant I, find $\sin 2A$.

We know from the Double Angle formula that $\sin 2A = 2 \sin A \cos A$ and we know that $\sin A = \dfrac{3}{5}$ so we can construct a right triangle with an acute angle A, with the side opposite A having length 3 and the hypotenuse having length 5.

Figure 1

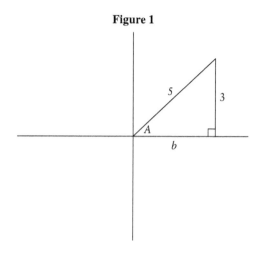

We can use the Pythagorean Theorem to find the other leg. We get: $b^2 + 3^2 = 5^2$, so $b = 4$.

Now we can plug into the formula: $\sin 2A = 2\left(\dfrac{3}{5}\right)\left(\dfrac{4}{5}\right) = \dfrac{24}{25}$.

By the way, if you use the Pythagorean Identity $\sin^2 A = 1 - \cos^2 A$, you can substitute this into the Double Angle formula for cosine and get: $\cos 2A = 2\cos^2 A - 1$. Similarly, if you use the identity, $\cos^2 A = 1 - \sin^2 A$, and substitute it into the Double Angle formula, you get: $\cos 2A = 1 - 2\sin^2 A$. All three of the cosine formulas are useful in different situations, but if you memorize the general formula, $\cos 2A = \cos^2 A - \sin^2 A$, you can easily derive the other two.

Of course there is a Double Angle formula for tangent as well. If we take the formula for the tangent of the sum of two angles, $\tan(A+B) = \dfrac{\tan A + \tan B}{1 - \tan A \tan B}$, and let $B = A$, we get: $\tan(A+A) = \dfrac{\tan A + \tan A}{1 - \tan A \tan A}$, which can be simplified to $\tan 2A = \dfrac{2\tan A}{1 - \tan^2 A}$.

Double Angle Formulas:

$$\sin 2A = 2\sin A \cos A$$

$$\cos 2A = \cos^2 A - \sin^2 A$$

$$\cos 2A = 2\cos^2 A - 1$$

$$\cos 2A = 1 - 2\sin^2 A$$

$$\tan 2A = \dfrac{2\tan A}{1 - \tan^2 A}$$

UNIT TWELVE: Trigonometric Angle Formulas

The Double Angle formulas are very useful for converting an expression that is difficult to work with into one that is easier to work with. For example, we can use them to reduce powers. If we have an expression with $\cos^2\theta$, we can reduce the power in the following way:
  Take the Double Angle formula for $\cos\theta$: $\cos 2\theta = 2\cos^2\theta - 1$.
  Add 1 to both sides and reverse the equation: $2\cos^2\theta = 1 + \cos 2\theta$.
  And divide by 2: $\cos^2\theta = \dfrac{1+\cos 2\theta}{2}$.
  Similarly, we can reduce $\sin^2\theta$ to $\sin^2\theta = \dfrac{1-\cos 2\theta}{2}$.
  If we want to reduce $\tan^2\theta$, we can divide the formula for $\sin^2\theta$ by the one for $\cos^2\theta$. We get: $\tan^2\theta = \dfrac{1-\cos 2\theta}{1+\cos 2\theta}$.
  Remember these. They will be very useful when you do Integral Calculus!

Power Reducing Formulas:

$$\sin^2\theta = \frac{1-\cos 2\theta}{2}$$

$$\cos^2\theta = \frac{1+\cos 2\theta}{2}$$

$$\tan^2\theta = \frac{1-\cos 2\theta}{1+\cos 2\theta}$$

Example 7: Rewrite $\cos^4\theta$ in terms of trigonometric functions of power 1.
First, we can rewrite this as the square of a square: $\cos^4\theta = \left(\cos^2\theta\right)^2$.
  Next, use the Power Reducing formula to substitute: $= \left(\dfrac{1+\cos 2\theta}{2}\right)^2$.
  Expand: $= \dfrac{1+2\cos 2\theta + \cos^2 2\theta}{4}$.
  We can break this into three expressions: $= \dfrac{1}{4} + \dfrac{2}{4}\cos 2\theta + \dfrac{1}{4}\cos^2 2\theta$.
  Now we can use the Power Reducing formula again on the last term:
$= \dfrac{1}{4} + \dfrac{2}{4}\cos 2\theta + \dfrac{1}{4}\left(\dfrac{1+\cos 4\theta}{2}\right)$.
  Which we can simplify to: $= \dfrac{1}{4} + \dfrac{2}{4}\cos 2\theta + \dfrac{1}{8} + \dfrac{\cos 4\theta}{8}$.
  So, $\cos^4\theta = \dfrac{3}{8} + \dfrac{1}{2}\cos 2\theta + \dfrac{1}{8}\cos 4\theta$.

There is one final set of formulas to learn. These are the Half-Angle formulas. We can get these by making a simple substitution into the Power Reducing formulas. Let $A = 2\theta$, which means that $\dfrac{A}{2} = \theta$. Substitute this into the formula for $\sin^2\theta$: $\sin^2\dfrac{A}{2} = \dfrac{1-\cos A}{2}$. Now take the square root of both sides: $\sin\dfrac{A}{2} = \pm\sqrt{\dfrac{1-\cos A}{2}}$. Notice that we need the $\pm$ sign because of the ambiguity of the square root.

In a similar fashion, we can get the Half-Angle formula for cosine and tangent: $\cos\dfrac{A}{2} = \pm\sqrt{\dfrac{1+\cos A}{2}}$ and $\tan\dfrac{A}{2} = \pm\sqrt{\dfrac{1-\cos A}{1+\cos A}}$.

Half-Angle Formulas:

$$\sin\dfrac{A}{2} = \pm\sqrt{\dfrac{1-\cos A}{2}}$$

$$\cos\dfrac{A}{2} = \pm\sqrt{\dfrac{1+\cos A}{2}}$$

$$\tan\dfrac{A}{2} = \pm\sqrt{\dfrac{1-\cos A}{1+\cos A}}$$

Example 8: Use the Half-Angle formula to find $\cos\dfrac{\pi}{8}$.

According to the formula: $\cos\dfrac{\pi}{8} = \sqrt{\dfrac{1+\cos\dfrac{\pi}{4}}{2}}$. We can get rid of the minus sign because we know that $\dfrac{\pi}{8}$ will be in Quadrant I, and therefore the cosine will be positive. Now we can substitute the value of $\cos\dfrac{\pi}{4} = \dfrac{\sqrt{2}}{2}$: $\cos\dfrac{\pi}{8} = \sqrt{\dfrac{1+\dfrac{\sqrt{2}}{2}}{2}}$, which can be simplified to $\cos\dfrac{\pi}{8} = \dfrac{\sqrt{2+\sqrt{2}}}{2}$.

Time to practice!

Practice Problems

Practice problem 1: Find the exact value of $\sin 50° \cos 20° - \cos 50° \sin 20°$.

Practice problem 2: Find the exact value of $\dfrac{\tan\dfrac{\pi}{5} - \tan\dfrac{\pi}{3}}{1 + \tan\dfrac{\pi}{5}\tan\dfrac{\pi}{3}}$.

Practice problem 3: If $\tan A = \dfrac{3}{7}$ and $\tan B = -\dfrac{2}{5}$, with A in quadrant I and B in quadrant II, find $\cos(A+B)$.

UNIT TWELVE: Trigonometric Angle Formulas

Practice problem 4: If $\sin A = \dfrac{4}{9}$ and $\cos B = \dfrac{3}{8}$, with A in quadrant I and B in quadrant IV, find $\tan(A - B)$.

Practice problem 5: Prove the following identity: $\sin(\theta - 90°) = -\cos\theta$.

Practice problem 6: Prove the following identity: $\dfrac{\sin(x+y)}{\sin(x-y)} = \dfrac{\tan x + \tan y}{\tan x - \tan y}$.

Practice problem 7: Rewrite $\sin 2\theta + \cos 2\theta$ in terms of $\sin\theta$ and $\cos\theta$.

Practice problem 8: Rewrite $\sin 3\theta + \cos 2\theta$ in terms of $\sin\theta$.

Practice problem 9: Prove the identity: $\sin(x-y) + \sin(x+y) = 2\sin x \cos y$.

Practice problem 10: Prove the identity: $\cos(x-y) + \cos(x+y) = 2\cos x \cos y$.

Practice problem 11: Find the exact solutions of $\sin 2x - \cos x = 0$ on the interval $0 \le x < 2\pi$.

Practice problem 12: Find the exact solutions of $\cos 2\theta + \sin\theta = 0$ on the interval $0° \le \theta < 360°$.

Practice problem 13: Use the Half-Angle formulas to find the exact value of $\sin 15°$.

Practice problem 14: Use the Half-Angle formulas to find the exact value of $\tan \dfrac{7\pi}{12}$.

Practice problem 15: Prove the following identity: $\cos^3\theta = \dfrac{1}{2}\cos\theta + \dfrac{1}{2}\cos\theta\cos 2\theta$.

Practice problem 16: Prove the following identity: $\tan\dfrac{\theta}{2} = \dfrac{\sin\theta}{1+\cos\theta}$.

Solutions to the Practice Problems

Solution to practice problem 1: *Find the exact value of* $\sin 50° \cos 20° - \cos 50° \sin 20°$.

Notice that this is the same form as $\sin(A - B) = \sin A \cos B - \cos A \sin B$, where $A = 50°$ and $B = 20°$. This means that we can rewrite $\sin 50° \cos 20° - \cos 50° \sin 20°$ as $\sin(50° - 20°)$. This reduces to $\sin 30° = \dfrac{1}{2}$.

Solution to practice problem 2: *Find the exact value of* $\dfrac{\tan\dfrac{\pi}{5}-\tan\dfrac{\pi}{3}}{1+\tan\dfrac{\pi}{5}\tan\dfrac{\pi}{3}}$.

Notice that this is the same form as $\tan(A-B)=\dfrac{\tan A-\tan B}{1+\tan A\tan B}$, where $A=\dfrac{\pi}{5}$ and $B=\dfrac{\pi}{3}$. This means that we can rewrite $\dfrac{\tan\dfrac{\pi}{5}-\tan\dfrac{\pi}{3}}{1+\tan\dfrac{\pi}{5}\tan\dfrac{\pi}{3}}$ as $\tan\left(\dfrac{\pi}{5}-\dfrac{\pi}{3}\right)$, which reduces to $\tan\left(-\dfrac{2\pi}{15}\right)$.

Solution to practice problem 3: *If* $\tan A=\dfrac{3}{7}$ *and* $\tan B=-\dfrac{2}{5}$, *with A in Quadrant I and B in Quadrant II, find* $\cos(A+B)$.

We can construct a triangle in Quadrant I with the side opposite angle A equal to 3 and the adjacent side equal to 7:

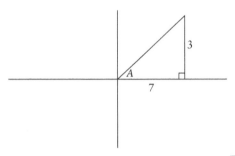

We can use the Pythagorean Theorem to find the hypotenuse: $\sqrt{3^2+7^2}=\sqrt{58}$.

Similarly, we can construct a triangle in quadrant II with the side opposite angle B equal to 2 and the adjacent side equal to −5:

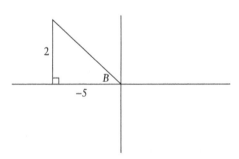

We can use the Pythagorean Theorem to find the hypotenuse: $\sqrt{2^2+5^2}=\sqrt{29}$.

UNIT TWELVE: Trigonometric Angle Formulas   137

Now we can use our triangles to find: $\sin A = \dfrac{3}{\sqrt{58}}$, $\cos A = \dfrac{7}{\sqrt{58}}$, $\sin B = \dfrac{2}{\sqrt{29}}$, and $\cos B = -\dfrac{5}{\sqrt{29}}$. Finally, we can plug the values into the Angle Addition formula for cosine: $\cos(A+B) = \left(\dfrac{7}{\sqrt{58}}\right)\left(-\dfrac{5}{\sqrt{29}}\right) - \left(\dfrac{3}{\sqrt{58}}\right)\left(\dfrac{2}{\sqrt{29}}\right) = -\dfrac{35}{29\sqrt{2}} - \dfrac{6}{29\sqrt{2}} = -\dfrac{41}{29\sqrt{2}}$.

Solution to practice problem 4: *If* $\sin A = \dfrac{4}{9}$ *and* $\cos B = \dfrac{3}{8}$, *with A in Quadrant I and B in Quadrant IV, find* $\tan(A - B)$.

We can construct a triangle in Quadrant I with the side opposite angle A equal to 4 and the hypotenuse equal to 9:

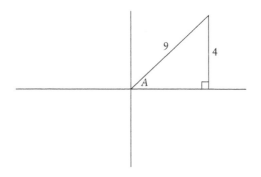

We can use the Pythagorean Theorem to find the other leg: $\sqrt{9^2 - 4^2} = \sqrt{65}$.

Similarly, we can construct a triangle in Quadrant IV with the side adjacent to angle B equal to 3 and the hypotenuse equal to 8:

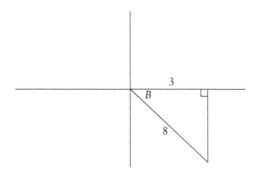

We can use the Pythagorean Theorem to find the other leg: $\sqrt{8^2 - 3^2} = \sqrt{55}$.

Now we can use our triangles to find: $\tan A = \dfrac{4}{\sqrt{65}}$ and $\tan B = \dfrac{\text{opp}}{\text{adj}} = -\dfrac{\sqrt{55}}{3}$.
Finally, we can plug the values into the Angle Difference formula for tangent:

$$\tan(A - B) = \dfrac{\dfrac{4}{\sqrt{65}} - \left(-\dfrac{\sqrt{55}}{3}\right)}{1 + \left(\dfrac{4}{\sqrt{65}}\right)\left(-\dfrac{\sqrt{55}}{3}\right)}.$$

Solution to practice problem 5: *Prove the following identity:* $\sin(\theta - 90°) = -\cos\theta$.
Let's use the Angle Difference formula for sine: $\sin(A - B) = \sin A \cos B - \cos A \sin B$ and plug in $A = \theta$ and $B = 90°$: $\sin(\theta - 90°) = \sin\theta\cos 90° - \cos\theta\sin 90°$. Next, we can plug in $\cos 90° = 0$ and $\sin 90° = 1$ to get: $\sin(\theta - 90°) = \sin\theta(0) - \cos\theta(1) = -\cos\theta$.

Solution to practice problem 6: *Prove the following identity*: $\dfrac{\sin(x+y)}{\sin(x-y)} = \dfrac{\tan x + \tan y}{\tan x - \tan y}$.

Let's work with the left side. We can use the Angle Addition formula and the Angle Difference formulas for sine: $\dfrac{\sin x \cos y + \cos x \sin y}{\sin x \cos y - \cos x \sin y} =$. Next, divide the terms in the numerator and the denominator by $\cos x \cos y$: $\dfrac{\dfrac{\sin x \cos y}{\cos x \cos y} + \dfrac{\cos x \sin y}{\cos x \cos y}}{\dfrac{\sin x \cos y}{\cos x \cos y} - \dfrac{\cos x \sin y}{\cos x \cos y}} =$. Finally, we can use the relationship $\tan\theta = \dfrac{\sin\theta}{\cos\theta}$ to simplify the expression: $\dfrac{\tan x + \tan y}{\tan x - \tan y} =$.

Solution to practice problem 7: *Rewrite* $\sin 2\theta + \cos 2\theta$ *in terms of* $\sin\theta$ *and* $\cos\theta$.
We can use the Double Angle formulas to get: $2\sin\theta\cos\theta + 1 - 2\sin^2\theta$.

Solution to practice problem 8: *Rewrite* $\sin 3\theta + \cos 2\theta$ *in terms of* $\sin\theta$.
We can use the Angle Addition formula to find $\sin 3\theta$: $\sin 3\theta = \sin(\theta + 2\theta) = \sin\theta\cos 2\theta + \sin 2\theta\cos\theta$. Next, we can use the Double Angle formulas and substitute: $\sin\theta(1 - 2\sin^2\theta) + (2\sin\theta\cos\theta)\cos\theta$. We can simplify this to: $\sin\theta - 2\sin^3\theta + 2\sin\theta\cos^2\theta$. Now we can substitute $\cos^2\theta = 1 - \sin^2\theta$: $\sin\theta - 2\sin^3\theta + 2\sin\theta(1 - \sin^2\theta)$. This simplifies to: $\sin\theta - 2\sin^3\theta + 2\sin\theta - 2\sin^3\theta = 3\sin\theta - 4\sin^3\theta$.
Now, we can add $\cos 2\theta = 1 - 2\sin^2\theta$ and get: $3\sin\theta - 4\sin^3\theta + 1 - 2\sin^2\theta$.

UNIT TWELVE: Trigonometric Angle Formulas 139

Solution to practice problem 9: *Prove the identity*: $\sin(x-y)+\sin(x+y)= 2\sin x\cos y$.
We can use the Angle Sum and Difference formulas to rewrite the left side as: $\sin x\cos y - \cos x\sin y + \sin x\cos y + \cos x\sin y =$.
And simplify: $2\sin x\cos y =$

Solution to practice problem 10: *Prove the identity*: $\cos(x-y)+\cos(x+y) = 2\cos x\cos y$.
We can use the Angle Sum and Difference formulas to rewrite the left side as: $\cos x\cos y + \sin x\sin y + \cos x\cos y - \sin x\sin y =$.
And simplify: $2\cos x\cos y =$

Solution to practice problem 11: *Find the exact solutions of* $\sin 2x - \cos x = 0$ *on the interval* $0 \le x < 2\pi$.
First, let's use the Double Angle formula to rewrite $\sin 2x$: $2\sin x\cos x - \cos x = 0$.
Next, factor out $\cos x$: $\cos x(2\sin x - 1) = 0$. This means that either $\cos x = 0$ or $2\sin x - 1 = 0$, or $\sin x = \frac{1}{2}$. $\cos x = 0$ when $x = \frac{\pi}{2}$ or $x = \frac{3\pi}{2}$ (on the interval). $\sin x = \frac{1}{2}$ when $x = \frac{\pi}{6}$ or $x = \frac{5\pi}{6}$.

Solution to practice problem 12: *Find the exact solutions of* $\cos 2\theta + \sin\theta = 0$ *on the interval* $0° \le \theta < 360°$.
First, let's use the Double Angle formula to rewrite $\cos 2\theta$: $1 - 2\sin^2\theta + \sin\theta = 0$. (Why did we choose that formula and not one of the other two? Because now we have the equation in terms of $\sin\theta$ only, which will make it simple to factor. Try the other two formulas to see that this is the best choice.)
Rewrite the equation: $2\sin^2\theta - \sin\theta - 1 = 0$ to make it easier to factor.
We get: $(2\sin\theta + 1)(\sin\theta - 1) = 0$.
This means that either $2\sin\theta + 1 = 0$, or $\sin\theta = -\frac{1}{2}$. Which means that $\sin\theta = 1$. $\sin\theta = -\frac{1}{2}$ when $\theta = 210°$ or $\theta = 330°$. $\sin\theta = 1$ when $\theta = 90°$.

Solution to practice problem 13: *Use the Half-Angle formulas to find an exact value of* $\sin 15°$.
Remember that the Half-Angle formula says that $\sin\frac{\theta}{2} = \pm\sqrt{\frac{1-\cos\theta}{2}}$. If we let $\theta = 30°$, we get: $\sin 15° = \sqrt{\frac{1-\cos 30°}{2}}$. (We know that this will be positive because $15°$ is in Quadrant I.)

We know that $\cos 30° = \dfrac{\sqrt{3}}{2}$: $\sin 15° = \sqrt{\dfrac{1-\dfrac{\sqrt{3}}{2}}{2}}$, which simplifies to $\sin 15° = \dfrac{\sqrt{2-\sqrt{3}}}{2}$.

Solution to practice problem 14: *Use the Half-Angle formulas to find the exact value of* $\tan \dfrac{7\pi}{12}$.

Remember that the Half-Angle formula says that $\tan \dfrac{x}{2} = \pm\sqrt{\dfrac{1-\cos x}{1+\cos x}}$. If we let $x = \dfrac{7\pi}{6}$, we get: $\tan \dfrac{7\pi}{12} = -\sqrt{\dfrac{1-\cos\dfrac{7\pi}{6}}{1+\cos\dfrac{7\pi}{6}}}$. (We know that this will be negative because $\dfrac{7\pi}{12}$ is in Quadrant II.)

We know that $\cos\dfrac{7\pi}{6} = -\dfrac{\sqrt{3}}{2}$: $\tan\dfrac{7\pi}{12} = -\sqrt{\dfrac{1+\dfrac{\sqrt{3}}{2}}{1-\dfrac{\sqrt{3}}{2}}}$, which simplifies to $\tan\dfrac{7\pi}{12} = -\sqrt{\dfrac{2+\sqrt{3}}{2-\sqrt{3}}}$.

Solution to practice problem 15: *Prove the following identity*: $\cos^3\theta = \dfrac{1}{2}\cos\theta + \dfrac{1}{2}\cos\theta\cos 2\theta$. Let's work with the right side. First, factor out $\dfrac{1}{2}\cos\theta$:

$= \dfrac{1}{2}\cos\theta(1+\cos 2\theta)$.

Next, use the Double Angle formula for cosine and substitute:

$= \dfrac{1}{2}\cos\theta\left(1+\left(2\cos^2\theta - 1\right)\right)$. (Why that one and not one of the other two? Because the left side is in terms of $\cos\theta$, so we want to avoid introducing a $\sin\theta$ term).

Simplify: $= \dfrac{1}{2}\cos\theta\left(2\cos^2\theta\right)$, which reduces to $= \cos^3\theta$.

UNIT TWELVE: Trigonometric Angle Formulas 141

Solution to practice problem 16: *Prove the following identity*: $\tan\dfrac{\theta}{2} = \dfrac{\sin\theta}{1+\cos\theta}$.

Remember the Half-Angle formula $\tan\dfrac{\theta}{2} = \pm\sqrt{\dfrac{1-\cos\theta}{1+\cos\theta}}$. Multiply the expression inside the radical by the conjugate of the numerator: $= \pm\sqrt{\dfrac{1-\cos\theta}{1+\cos\theta}\dfrac{1+\cos\theta}{1+\cos\theta}}$.

This simplifies to: $= \pm\sqrt{\dfrac{1-\cos^2\theta}{(1+\cos\theta)^2}}$.

Next, remember that $\sin^2\theta = 1 - \cos^2\theta$ and substitute: $= \pm\sqrt{\dfrac{\sin^2\theta}{(1+\cos\theta)^2}}$. Now take the square root: $= \dfrac{\sin\theta}{1+\cos\theta}$. Note that we can ignore the $\pm$ because $\sin\theta$ and $\tan\dfrac{\theta}{2}$ always have the same sign. (Graph them and you will see that $1 \pm \cos\theta$ is never negative.)

UNIT THIRTEEN
The Law of Sines

Up until now, all of the Trigonometry that we have learned has involved right triangles. Now we are going to learn how to use trig when we are working with any type of triangle. Of course, all of the identities and relationships still apply, and the values of the trig functions of an angle don't change. What we are going to learn is how to find the missing sides and angles of a triangle given some of the parts. This is called *solving the triangle*. One of the ways that we can do so is with the Law of Sines. (Another is the Law of Cosines, which we will explore after this.)

By the way, so far we have not used a calculator to find the values of trig functions (although we certainly could have). Here, we will be using angles where we will need to use a calculator to find their sines, cosines, etc.

Suppose we have a triangle with angles A, B, and C, and that the sides opposite those angles are a, b, and c, respectively:

Figure 1

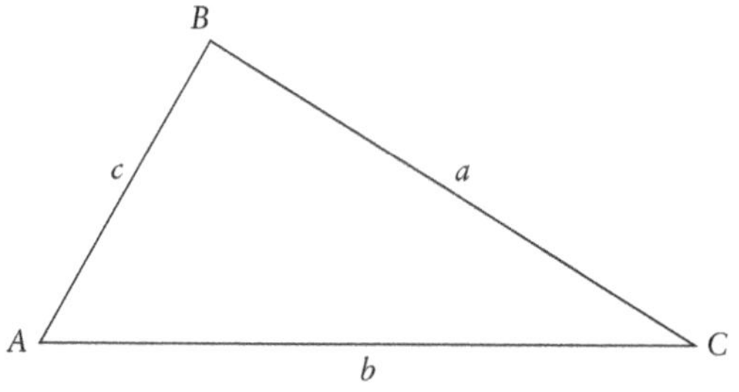

The Law of Sines states that the ratio of the sine of any angle to its opposite side equals the ratio of the sine of one of the other angles to its opposite side. In other words,

$$\frac{\sin A}{a} = \frac{\sin B}{b} = \frac{\sin C}{c}.$$

By the way, we could use the Law of Sines with right triangles, but it is much easier to use the SOHCAHTOA relationships instead.

We will use the Law of Sines to solve a triangle when we are given *AAS* or *ASA*, as well as for SSA. If we are given different parts of a triangle, we will instead use the Law of Cosines.

A brief note about the sketches in this section. We are going to sketch triangles to help us organize the information that we are given. The triangles are **not** *drawn to scale and may, in fact, be drawn as acute when they are actually obtuse, or when there is no possible triangle. Therefore, do* **not** *rely on the sketch as anything other than a tool to help us set up the equations.*

UNIT THIRTEEN: The Law of Sines

Let's do an example.

Example 1: Given $\angle A = 38°$, $\angle B = 46°$, and $a = 12$, solve $\triangle ABC$.

Let's make a sketch and put in the information that we have been given:

Figure 2

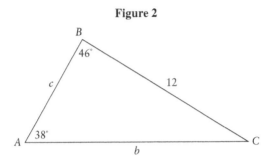

We are given AAS and we can easily find $\angle C = 180° - 38° - 46° = 96°$. Now to find sides b and c.

Let's use the Law of Sines to find b: $\dfrac{\sin 38°}{12} = \dfrac{\sin 46°}{b}$.

A little algebra gives us: $b = \dfrac{12 \sin 46°}{\sin 38°}$. We can use the calculator to solve for b: $b \approx 14.021$.

Let's use the Law of Sines again to find c: $\dfrac{\sin 38°}{12} = \dfrac{\sin 96°}{c}$.

Once again, we can use a little algebra: $c = \dfrac{12 \sin 96°}{\sin 38°}$, and, therefore, $c \approx 19.384$.

Notice that we did not use the ratio $\dfrac{\sin B}{b}$ to find c. Although we could have, we don't know b precisely and we do know a, so using the ratio $\dfrac{\sin A}{a}$ is preferable.

The solved triangle is therefore:

$$\angle A = 38°, \angle B = 46°, \angle C = 96°;$$

$$a = 12, b \approx 14.021, c \approx 19.384.$$

Let's do another example.

Example 2: Given $\angle A = 50°$, $\angle C = 28°$, and $b = 10$, solve $\triangle ABC$.

Let's make a sketch and put in the information that we have been given:

Figure 3

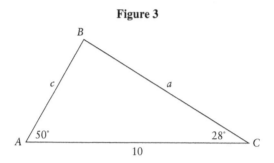

Here, we are given *ASA*. If we want to use the Law of Sines, we are first going to need to find $\angle B$. Otherwise, we will have a ratio where we don't know the numerator or the denominator. We can easily find $\angle B = 180° - 28° - 50° = 102°$. Now to find sides a and c.

Let's use the Law of Sines to find a: $\dfrac{\sin 50°}{a} = \dfrac{\sin 102°}{10}$.

A little algebra gives us: $a = \dfrac{10 \sin 50°}{\sin 102°}$. We can use the calculator to solve for a: $a \approx 7.832$.

Let's use the Law of Sines again to find c: $\dfrac{\sin 102°}{10} = \dfrac{\sin 28°}{c}$.

Once again, we can use a little algebra: $c = \dfrac{10 \sin 28°}{\sin 102°}$, and, therefore, $c \approx 4.800$.

Notice that we did not use the ratio $\dfrac{\sin A}{a}$ to find c. As with Example 1, we could have used that ratio, but we don't know a precisely and we do know b precisely. So, using the ratio $\dfrac{\sin B}{b}$ is preferable.

The solved triangle is, therefore:

$$\angle A = 50°, \angle B = 102°, \angle C = 28°;$$

$$a \approx 7.832, b = 10, c \approx 4.800.$$

If we are given two angles and a side of a triangle, we can always solve the triangle precisely. *Unfortunately*, the same is not true if we are given *SSA*. (If we have *SSS* or *SAS*, we will use the Law of Cosines). *When we have SSA, it is possible*

UNIT THIRTEEN: The Law of Sines   145

that there are zero, one, or two triangles. How will we know? Let's learn through some examples.

Example 3: Suppose we are given $\angle A = 32°$, $a = 7$, and $b = 9$, solve $\triangle ABC$.
First, let's make a sketch:

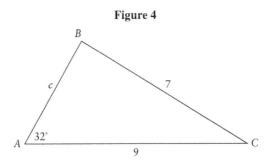

Figure 4

We are given *SSA*, so we will need to be alert for the possibility of two triangles.

Let's use the Law of Sines to find $\angle B$: $\dfrac{\sin 32°}{7} = \dfrac{\sin B}{9}$.

Multiply across by 9: $\dfrac{9\sin 32°}{7} = \sin B$.

So, $\sin B \approx .6813$, and $\sin^{-1}(0.6813) = B \approx 43°$.

Now here's where the ambiguity arises. Remember that the sine of an angle in Quadrant II is positive, so it's possible that $\angle B$ is also $180° - 43° = 137°$.

Let's solve for $\angle C$ in both cases. For the first case, $C = 180° - 32° - 43° = 105°$. For the second case, $C = 180° - 32° - 137° = 11°$.

Now we can find side c for both triangles. For the first one, $\dfrac{\sin 32°}{7} = \dfrac{\sin 105°}{c}$, and $c \approx 12.759$. For the second triangle, $\dfrac{\sin 32°}{7} = \dfrac{\sin 11°}{c}$, and $c \approx 2.521$. Which one is correct? They both are! This is why we call such a situation the *Ambiguous Case*.

There are two solved triangles:

$\angle A = 32°$, $\angle B \approx 43°$, $\angle C = 105°$; $\angle A = 32°$, $\angle B \approx 43°$, $\angle C = 11°$;

$a = 7, b = 9, c \approx 12.759$ $a = 7, b = 9, c \approx 2.521$

Now let's solve a slightly different triangle.

Example 4: $\angle A = 32°$, $a = 9$, and $b = 7$, solve $\triangle ABC$.
Notice that the only difference between this example and the previous one is that $b > a$ in the former and $a > b$ in the latter. This is a crucial difference!

Let's make a sketch:

Figure 5

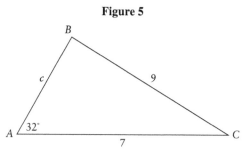

As in the previous example, we are given SSA, so we will need to be alert for the possibility of two triangles.

Let's use the Law of Sines to find $\angle B$: $\dfrac{\sin 32°}{9} = \dfrac{\sin B}{7}$.

Multiply across by 7: $\dfrac{7 \sin 32°}{9} = \sin B$.

So, $\sin B \approx .4122$ and $\sin^{-1}(0.4122) = B \approx 24°$.

Again, remember that the sine of an angle in Quadrant II is positive, so it's possible that $\angle B$ is also $180° - 24° = 156°$.

Let's solve for $\angle C$ in both cases. For the first case, $C = 180° - 32° - 24° = 124°$. For the second case, $C = 180° - 32° - 156° = -8°$. Obviously, the second case is impossible, so there is only one triangle.

Now we can find side c: $\dfrac{\sin 32°}{9} = \dfrac{\sin 124°}{c}$ and $c \approx 14.080$.

The solved triangle is:

$$\angle A = 32°, \angle B \approx 24°, \angle C = 124°;$$

$$a = 9, b = 7, c \approx 14.080.$$

Now, let's look at one more case.

Example 5: Suppose we are given $\angle A = 52°$, $a = 7$, and $b = 11$, solve $\triangle ABC$.

Let's make a sketch:

Figure 6

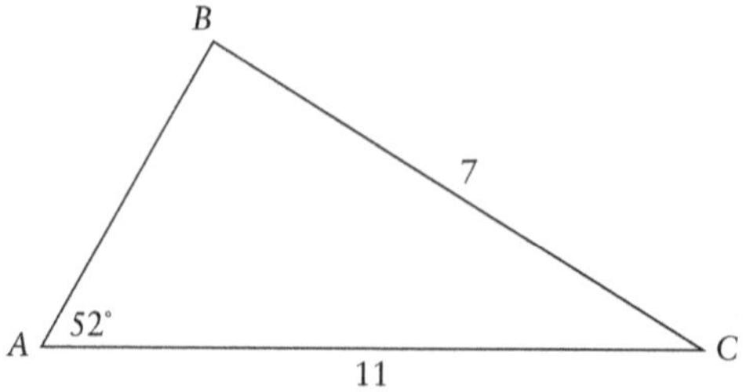

UNIT THIRTEEN: The Law of Sines

As in the previous example, we are given SSA, so we will need to be alert for the possibility of two triangles.

Let's use the Law of Sines to find $\angle B$: $\dfrac{\sin 52°}{7} = \dfrac{\sin B}{11}$.

Multiply across by 11: $\dfrac{11 \sin 52°}{7} = \sin B$.

So, $\sin B \approx 1.2383$. But this is not possible! The value of sine cannot be bigger than 1. Therefore, this is not a valid case, and no triangle is possible.

Now we have seen all three scenarios when we are given SSA. There could be zero, one, or two triangles. How will we know what to do? One way is to do what we just did. Apply the Law of Sines to solve for a missing angle. There are three possibilities:

> If we get two possible values for the third angle, then there are two triangles.
> If we only get one possible value, then there is no ambiguity and there is only one triangle.
> If we get a value greater than 1, then there is no triangle.

Let's do some practice problems!

Practice Problems

Practice problem 1: Solve the triangle: $\angle A = 42°$, $\angle B = 36°$, $b = 11$.
Practice problem 2: Solve the triangle: $\angle A = 51°$, $\angle B = 61°$, $a = 5$.
Practice problem 3: Solve the triangle: $\angle A = 32°$, $a = 3$, $b = 4$.
Practice problem 4: Solve the triangle: $\angle B = 70°$, $b = 13$, $c = 9$.
Practice problem 5: Solve the triangle: $\angle A = 53°$, $a = 14$, $b = 16$.
Practice problem 6: Solve the triangle: $\angle B = 38°$, $b = 23$, $c = 26$.
Practice problem 7: Solve the triangle: $\angle A = 36°$, $a = 3$, $b = 7$.
Practice problem 8: Solve the triangle: $\angle A = 46°$, $a = 8$, $b = 5$.
Practice problem 9: The distance between two lighthouses is 12 mi, and both observe a ship off shore. The angle of observation from Lighthouse A to the ship is $43°$, and the angle of observation from Lighthouse B to the ship is $71°$. What is the distance from the ship to each of the lighthouses?
Practice problem 10: Two flags, A and B, are situated 80 ft apart on one side of a river. An observer, C, stands on the other side of the river and observes that $\angle CBA = 48°$ and $\angle ACB = 78°$. How wide is the river?

Solutions to the Practice Problems

Solution to practice problem 1: *Solve the triangle: $\angle A = 42°$, $\angle B = 36°$, $b = 11$.*
First, let's draw a triangle and put in the information that we are given:

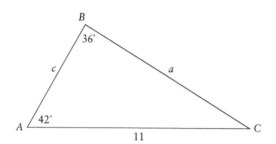

We can easily find angle C because the sum of the angles of a triangle is always $180°$: $C = 180° - 42° - 36° = 102°$. Next, we can find side a using the Law of Sines: $\dfrac{\sin 36°}{11} = \dfrac{\sin 42°}{a}$.

Use a little algebra to isolate a: $a = \dfrac{11 \sin 42°}{\sin 36°} \approx 12.522$. Finally, we can find c, again using the Law of Sines: $\dfrac{\sin 36°}{11} = \dfrac{\sin 102°}{c}$. A little algebra gives: $c = \dfrac{11 \sin 102°}{\sin 36°} \approx 18.305$.

The solved triangle is, therefore:

$$\angle A = 42°, \angle B = 36°, \angle C = 102°;$$
$$a \approx 12.522, b = 11, c \approx 18.305.$$

Solution to practice problem 2: *Solve the triangle: $\angle A = 51°$, $\angle B = 61°$, $a = 5$.*
First, let's draw a triangle and put in the information that we are given.

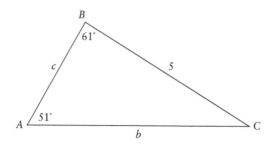

UNIT THIRTEEN: The Law of Sines

We can easily find angle C because the sum of the angles of a triangle is always $180°$: $C = 180° - 51° - 61° = 68°$. Next, we can find side b using the Law of Sines: $\dfrac{\sin 51°}{5} = \dfrac{\sin 61°}{b}$.

Use a little algebra to isolate b: $b = \dfrac{5\sin 61°}{\sin 51°} \approx 5.627$. Finally, we can find c, again using the Law of Sines: $\dfrac{\sin 51°}{5} = \dfrac{\sin 68°}{c}$. A little algebra gives: $c = \dfrac{5\sin 68°}{\sin 51°} \approx 5.965$.

The solved triangle is, therefore:

$$\angle A = 51°, \angle B = 61°, \angle C = 68°;$$

$$a = 5, b \approx 5.627, c \approx 5.965.$$

Solution to practice problem 3: *Solve the triangle:* $\angle A = 32°$, $a = 3$, $b = 4$.

First, let's draw a triangle and put in the information that we are given;

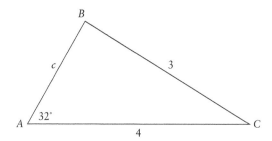

Let's find angle B using the Law of Sines: $\dfrac{\sin 32°}{3} = \dfrac{\sin B}{4}$. Multiply through by 4: $\dfrac{4\sin 32°}{3} = \sin B$. So, $\sin B \approx 0.7066$, and $B = \sin^{-1}(0.7066) \approx 45°$.

Next, we can find angle C because the sum of the angles of a triangle is always $180°$: $C = 180° - 32° - 45° = 103°$.

Finally, we can find side c using the Law of Sines: $\dfrac{\sin 32°}{3} = \dfrac{\sin 103°}{c}$. Use a little algebra to isolate c: $c = \dfrac{3\sin 103°}{\sin 32°} \approx 5.516$.

The solved triangle is, therefore:

$$\angle A = 32°, \angle B = 45°, \angle C = 103°;$$

$$a = 3, b = 4, c \approx 5.516.$$

Solution to practice problem 4: *Solve the triangle:* $\angle B = 70°, b = 13, c = 9$.
First, let's draw a triangle and put in the information that we are given:

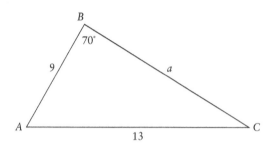

Let's find angle C using the Law of Sines: $\dfrac{\sin 70°}{13} = \dfrac{\sin C}{9}$. Multiply through by 9: $\dfrac{9\sin 70°}{13} = \sin C$. So, $\sin C \approx .6506$, and $C = \sin^{-1}(0.6506) \approx 41°$.

Next, we can find angle A because the sum of the angles of a triangle is always $180°$: $A = 180° - 70° - 41° = 69°$.

Finally, we can find side a using the Law of Sines: $\dfrac{\sin 70°}{13} = \dfrac{\sin 69°}{a}$. Use a little algebra to isolate a: $a = \dfrac{13\sin 69°}{\sin 70°} \approx 12.915$.

The solved triangle is, therefore:

$$\angle A = 69°, \angle B = 70°, \angle C = 41°;$$

$$a \approx 12.915, b = 13, c = 9.$$

Solution to practice problem 5: *Solve the triangle:* $\angle A = 53°, a = 14, b = 16$.
First, let's draw a triangle and put in the information that we are given:

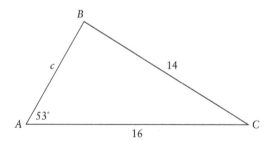

UNIT THIRTEEN: The Law of Sines

We are given *SSA*, so we will need to be alert for the possibility of two triangles.

Let's use the Law of Sines to find $\angle B$: $\dfrac{\sin 53°}{14} = \dfrac{\sin B}{16}$.

Multiply across by 16: $\dfrac{16 \sin 53°}{14} = \sin B$.

So, $\sin B \approx 0.9127$ and $\sin^{-1}(0.9127) = B \approx 66°$.

Remember that it's possible that $\angle B$ is also $180° - 66° = 114°$.

Let's solve for $\angle C$ in both cases. For the first case, $C = 180° - 53° - 66° = 61°$. For the second case, $C = 180° - 53° - 114° = 13°$.

Now we can find side c for both triangles. For the first one, $\dfrac{\sin 53°}{14} = \dfrac{\sin 61°}{c}$ and $c \approx 15.332$. For the second triangle, $\dfrac{\sin 53°}{14} = \dfrac{\sin 13°}{c}$ and $c \approx 3.943$. This is an example of the *Ambiguous Case*.

The two solved triangles are:

$$\angle A = 53° \quad \angle B = 66° \quad \angle C = 61°; \quad \angle A = 53° \quad \angle B = 114° \quad \angle C = 13°$$
$$a = 14 \quad b = 16 \quad c \approx 15.332 \quad\quad a = 14 \quad b = 16 \quad c \approx 3.943$$

Solution to practice problem 6: *Solve the triangle:* $\angle B = 38°$, $b = 23$, $c = 26$.

First, let's draw a triangle and put in the information that we are given:

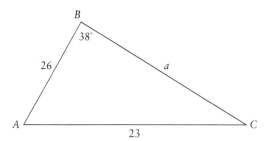

We are given *SSA*, so we will need to be alert for the possibility of two triangles.

Let's use the Law of Sines to find $\angle C$: $\dfrac{\sin 38°}{23} = \dfrac{\sin C}{26}$.

Multiply across by 26: $\dfrac{26 \sin 38°}{23} = \sin C$.

So $\sin C \approx 0.6960$ and $\sin^{-1}(0.6960) = C \approx 44°$.

Remember that it's possible that $\angle C$ is also $180° - 44° = 136°$.

Let's solve for $\angle A$ in both cases. For the first case, $A = 180° - 38° - 44° = 98°$. For the second case, $A = 180° - 38° - 136° = 6°$.

Now we can find side a for both triangles. For the first one, $\dfrac{\sin 38°}{23} = \dfrac{\sin 98°}{a}$ and $a \approx 36.995$. For the second triangle, $\dfrac{\sin 38°}{23} = \dfrac{\sin 6°}{a}$ and $a \approx 3.905$. This is an example of the *Ambiguous Case*.

The two solved triangles are:

$$\angle A = 98°, \angle B = 38°, \angle C = 44°; \quad \angle A = 6° \; \angle B = 38° \; \angle C = 136°$$

$$a \approx 36.995 \; b = 23 \; c = 26 \quad a \approx 3.905 \; b = 23 \; c = 26$$

Solution to practice problem 7: *Solve the triangle:* $\angle A = 36°, a = 3, b = 7.$
First, let's draw a triangle and put in the information that we are given:

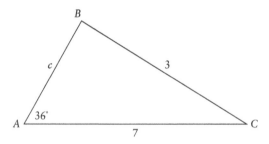

We are given SSA, so we will need to be alert for the possibility of two triangles.

Let's use the Law of Sines to find $\angle B$: $\dfrac{\sin 36°}{3} = \dfrac{\sin B}{7}$.

Multiply across by 7: $\dfrac{7 \sin 36°}{3} = \sin B$.

So, $\sin B \approx 1.3715$. The value of sine cannot be larger than 1, so this is not a valid case. There is no triangle possible.

Solution to practice problem 8: *Solve the triangle:* $\angle A = 46°, a = 8, b = 5.$
First, let's draw a triangle and put in the information that we are given:

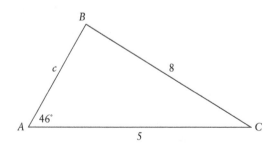

UNIT THIRTEEN: The Law of Sines 153

We are given SSA, so we will need to be alert for the possibility of two triangles.

Let's use the Law of Sines to find $\angle B$: $\dfrac{\sin 46°}{8} = \dfrac{\sin B}{5}$.

Multiply across by 5: $\dfrac{5\sin 46°}{8} = \sin B$.

So, $\sin B \approx 0.4496$ and $\sin^{-1}(0.4496) = B \approx 27°$.

Again, remember that the sine of an angle in Quadrant II is positive, so it's possible that $\angle B$ is also $180° - 27° = 153°$.

Let's solve for $\angle C$ in both cases. For the first case, $C = 180° - 46° - 27° = 107°$. For the second case, $C = 180° - 46° - 153° = -19°$. Obviously, the second case is impossible, so this time there is only one triangle.

Now we can find side c: $\dfrac{\sin 46°}{8} = \dfrac{\sin 107°}{c}$ and $c \approx 10.635$.

The solved triangle is:

$\angle A = 32°, \angle B \approx 27°, \angle C = 107°$;

$a = 8, b = 5, c \approx 10.635$.

Solution to practice problem 9: *The distance between two lighthouses is 12 mi, and both observe a ship off shore. The angle of observation from Lighthouse A to the ship is 43°, and the angle of observation from Lighthouse B to the ship is 71°. What is the distance from the ship to each of the lighthouses?*

First, let's make a sketch and put in the information that we are given:

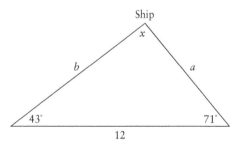

Let's find the missing angle, which we will call x: $x = 180° - 43° - 71° = 66°$.

Next, let's solve for a, the distance from Lighthouse A to the ship: $\dfrac{\sin 43°}{a} = \dfrac{\sin 66°}{12}$ and

$$a \approx \dfrac{12\sin 43°}{\sin 66°} \approx 8.958 \text{ mi.}$$

Finally, let's solve for B, the distance from Lighthouse B to the ship: $\frac{\sin 71°}{b} = \frac{\sin 66°}{12}$ and

$$b \approx \frac{12 \sin 71°}{\sin 66°} \approx 12.420 \text{ mi}.$$

Solution to practice problem 10: *Two flags, A and B, are situated 80 ft apart on one side of a river. An observer, C, stands on the other side of the river and observes that $\angle CBA = 48°$ and $\angle ACB = 78°$. How wide is the river?*

First, let's make a sketch and put in the information that we are given:

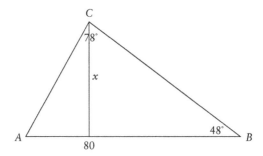

We are looking for x, the width of the river. Let's start by finding the missing angle: $A = 180° - 48° - 78° = 54°$.

Next, let's use the Law of Sines to find side BC: $\frac{\sin 78°}{80} = \frac{\sin 54°}{BC}$. With a little algebra, we find that $BC = \frac{80 \sin 54°}{\sin 78°} \approx 66.167$ ft. Now, we can find x using right triangle trigonometry: $\sin 48° = \frac{x}{66.167}$ and $x \approx 49.172$ ft.

UNIT FOURTEEN
The Law of Cosines and Area Formulas

Now we are going to learn another formula to help you solve non-right triangles, the Law of Cosines. As you will see, the Law of Cosines is similar to the Pythagorean Theorem with an extra term. In fact, the Law of Cosines, when applied to a right triangle, becomes the Pythagorean Theorem.

Suppose we have the following triangle, which is the same general triangle that we use for the Law of Sines:

Figure 1

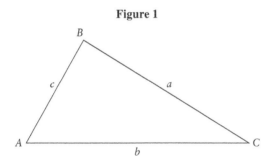

The Law of Cosines says that:

$$c^2 = a^2 + b^2 - 2ab\cos C$$
$$a^2 = b^2 + c^2 - 2bc\cos A$$
$$b^2 = a^2 + c^2 - 2ac\cos B$$

Notice that these are really all the same. We present it this way so that you can see that it doesn't matter which letters you use; it is the relationship that matters.

How will we know whether to use the Law of Sines or the Law of Cosines?
If we are given *ASA*, *SAA*, or *SSA*, we use the Law of Sines.
If we are given *SAS* or *SSS*, we use the Law of Cosines.
Let's use the Law of Cosines in an example.

Example 1: Solve triangle *ABC*, given $a = 12$, $b = 6$, and $C = 35°$.

First, let's make a sketch:

Figure 2

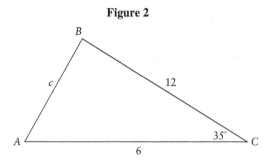

Notice that we are given *SAS*, so we will use the Law of Cosines. Let's solve for *c*:

$$c^2 = 12^2 + 6^2 - 2(12)(6)\cos 35°$$

$$c^2 = 144 + 36 - 144(.8192)$$

$$c^2 = 180 - 117.9648 = 62.0352$$

$$c \approx 7.88.$$

Next, we need to find one of the two unknown angles. Although we could use either the Law of Sines or the Law of Cosines to find one of the angles, it is generally better to use the Law of Cosines because the Arccosine function distinguishes between acute and obtuse angles, whereas the Arcsine function does not.

Let's find *A*: $12^2 = 6^2 + 7.88^2 - 2(6)(7.88)\cos A$

$$144 = 36 + 62.0944 - 94.56\cos A$$

$$144 = 98.0944 - 94.56\cos A$$

$$45.9056 = -94.56\cos A$$

$$-\frac{45.9056}{94.56} = \cos A$$

$$A = \cos^{-1}\left(-\frac{45.9056}{94.56}\right) \approx 119°.$$

Finally, we can find *B* because the sum of the angles of a triangle is always 180°:

$$B = 180° - 119° - 35° = 26°.$$

UNIT FOURTEEN: The Law of Cosines and Area Formulas

Therefore, the solved triangle is:

$$\angle A \approx 119°, \angle B \approx 26°, \angle C = 35°;$$

$$a = 12, b = 6, c \approx 7.88.$$

Let's do another example.

Example 2: Solve triangle ABC, given $a = 10$, $b = 13$, and $c = 8$.
First, let's make a sketch:

Figure 3

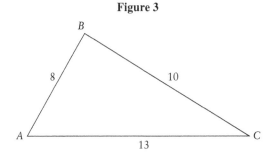

Notice that we are given SSS, so we will use the Law of Cosines. We will need to solve for two angles and then subtract the sum of the two angles from $180°$ to find the third angle.

Let's solve for C:

$$8^2 = 10^2 + 13^2 - 2(10)(13)\cos C$$

$$64 = 269 - 260\cos C$$

$$-205 = -260\cos C$$

$$\frac{205}{260} = \cos C$$

$$C = \cos^{-1}\left(\frac{205}{260}\right) \approx 38°.$$

Now, let's solve for B:

$$13^2 = 10^2 + 8^2 - 2(10)(8)\cos B$$

$$169 = 164 - 160\cos B$$

$$5 = -160 \cos B$$

$$-\frac{5}{160} = \cos B$$

$$B = \cos^{-1}\left(-\frac{5}{160}\right) \approx 92°.$$

Finally, we can find A because the sum of the angles of a triangle is always $180°$:

$$A = 180° - 92° - 38° = 50°.$$

Therefore, the solved triangle is:

$$\angle A \approx 50°, \ \angle B \approx 92°, \ \angle C \approx 38°;$$

$$a = 10, \ b = 13, \ c = 8.$$

We have one last formula to learn. This one will enable us to find the area of a triangle using Trigonometry. If we are given *SAS*, then the formula for area is:

$$\text{Area} = \frac{1}{2}ab\sin C = \frac{1}{2}ac\sin B = \frac{1}{2}bc\sin A.$$

Note that these are all really the same formula. They just use different sides and angles, depending on what we have been given.

Let's do an example.

Example 3: Find the area of the triangle with $a = 14$, $b = 12$, and $C = 65°$.
First, let's make a sketch:

Figure 4

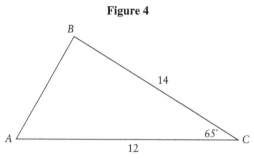

We are given *SAS*, so we can use the formula to find the area: $\text{Area} = \frac{1}{2}(14)(12)\sin 65° \approx 76.13$.

How about another example?

UNIT FOURTEEN: The Law of Cosines and Area Formulas

Example 4: Find the area of a regular pentagon inscribed in a circle of radius 10 inches.

First, let's make a sketch:

Figure 5

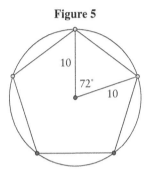

Notice that we can divide the pentagon into five congruent triangles. Each central angle will be $\dfrac{360°}{5} = 72°$, and the radii will form congruent sides of length 10.

We can use the area formula to find the area of one of the triangles and then multiply by 5 to find the area of the pentagon:

$$A = \frac{1}{2}(10)(10)\sin 72° \approx 47.553$$

$$(5)(47.553) \approx 237.76 \text{ in.}^2$$

By the way, because a triangle can always be thought of as half a parallelogram, we can use the area formula to find the area of a parallelogram by simply multiplying the formula by 2:

Figure 6

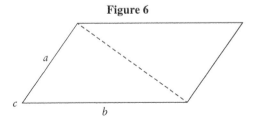

Area = $ab \sin C$

This formula will come in handy when you study vectors, among other topics. Are you ready for some practice problems?

Practice Problems

Practice problem 1: Solve the triangle: $\angle A = 62°$, $b = 19$, $c = 17$.
Practice problem 2: Solve the triangle: $\angle B = 39°$, $a = 16$, $c = 22$.
Practice problem 3: Solve the triangle: $\angle C = 105°$, $a = 5$, $b = 4$.
Practice problem 4: Solve the triangle: $a = 10$, $b = 17$, $c = 14$.
Practice problem 5: Solve the triangle: $a = 6$, $b = 8$, $c = 12$.
Practice problem 6: Find the area of the triangle: $\angle B = 58°$, $a = 27$ cm, $c = 19$ cm.
Practice problem 7: Find the area of the triangle: $\angle A = 101°$, $b = 10$ ft, $c = 20$ ft.
Practice problem 8: A parallelogram has sides of length 15 in. and 19 in. If the angle between the sides is $88°$, find the area of the parallelogram.
Practice problem 9: A surveyor stands some distance away from a pond. The surveyor measures the distance to one edge of the pond as 105 ft. She then turns through $50°$ and measures the distance to the other edge of the pond as 122 ft. Approximately how wide is the pond?
Practice problem 10: Find the area of a regular octagon (8 sides) inscribed in a circle of radius 8 cm.

Solutions to the Practice Problems

Solution to practice problem 1: *Solve the triangle:* $\angle A = 62°$, $b = 19$, $c = 17$.
First, let's draw a triangle and put in the information that we are given:

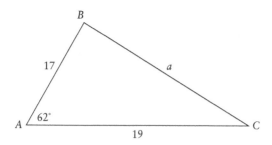

Notice that we are given *SAS*, so we will use the Law of Cosines. Let's solve for a:

$$a^2 = 19^2 + 17^2 - 2(19)(17)\cos 62°$$

$$a^2 = 650 - 646\cos 62°$$

$$a^2 \approx 346.7214$$

$$a \approx 18.62.$$

UNIT FOURTEEN: The Law of Cosines and Area Formulas

Next, let's find B:

$$19^2 = 17^2 + 18.62^2 - 2(17)(18.62)\cos B$$

$$361 = 635.7044 - 633.08\cos B$$

$$-274.7044 = -633.08\cos B$$

$$\frac{274.7044}{633.08} = \cos B$$

$$B = \cos^{-1}\left(\frac{274.7044}{633.08}\right) \approx 64°.$$

Finally, we can find C because the sum of the angles of a triangle is always $180°$:

$$C = 180° - 64° - 62° = 54°.$$

The solved triangle is, therefore:

$$\angle A = 62°,\ \angle B \approx 64°,\ \angle C \approx 54°;$$
$$a \approx 18.62,\ b = 19,\ c = 17.$$

Solution to practice problem 2: *Solve the triangle:* $\angle B = 39°$, $a = 16$, $c = 22$.
First, let's draw a triangle and put in the information that we are given:

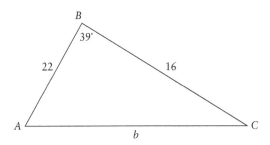

Notice that we are given *SAS*, so we will use the Law of Cosines. Let's solve for b:

$$b^2 = 22^2 + 16^2 - 2(22)(16)\cos 39°$$

$$b^2 = 740 - 704\cos 39°$$

$$b^2 \approx 192.8892$$

$$b \approx 13.89.$$

Next let's find A:

$$16^2 = 22^2 + 13.89^2 - 2(22)(13.89)\cos A$$

$$256 = 676.9321 - 611.16\cos A$$

$$-420.9321 = -611.16\cos A$$

$$\frac{420.9321}{611.16} = \cos A$$

$$A = \cos^{-1}\left(\frac{420.9321}{611.16}\right) \approx 46°.$$

Finally, we can find C because the sum of the angles of a triangle is always $180°$:

$$C = 180° - 46° - 39° = 95°.$$

The solved triangle is, therefore:

$$\angle A \approx 46°, \angle B = 39°, \angle C \approx 95°;$$

$$a = 16, b \approx 13.89, c = 22.$$

Solution to practice problem 3: *Solve the triangle:* $\angle C = 105°$, $a = 5$, $b = 4$.
First, let's draw a triangle and put in the information that we are given:

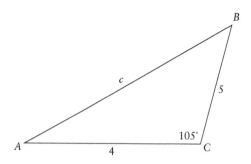

Let's use the Law of Cosines to solve for c:

$$c^2 = 5^2 + 4^2 - 2(5)(4)\cos 105°$$

$$c^2 = 41 - 40\cos 105°$$

$$c^2 \approx 51.3528$$

$$c \approx 7.17.$$

UNIT FOURTEEN: The Law of Cosines and Area Formulas

Next let's find A:

$$5^2 = 4^2 + 7.17^2 - 2(4)(7.17)\cos A$$

$$25 = 67.4089 - 57.36\cos A$$

$$-42.4089 = -57.36\cos A$$

$$\frac{42.4089}{57.36} = \cos A$$

$$A = \cos^{-1}\left(\frac{42.4089}{57.36}\right) \approx 42°.$$

Finally, we can find B because the sum of the angles of a triangle is always $180°$:

$$B = 180° - 42° - 105° = 33°.$$

The solved triangle is, therefore:

$$\angle A \approx 42°, \angle B \approx 33°, \angle C = 105°;$$
$$a = 5, \ b = 4, \ c \approx 7.17.$$

Solution to practice problem 4: *Solve the triangle:* $a = 10, b = 17, c = 14$.
First, let's draw a triangle and put in the information that we are given:

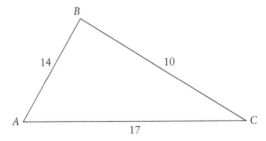

We are given SSS, so we will use the Law of Cosines. We will need to solve for two angles and can then subtract from $180°$ to find the third angle.
Let's solve for C:

$$14^2 = 10^2 + 17^2 - 2(10)(17)\cos C$$

$$196 = 389 - 340\cos C$$

$$-193 = -340\cos C$$

$$\frac{193}{340} = \cos C$$

$$C = \cos^{-1}\left(\frac{193}{340}\right) \approx 55°.$$

Now let's solve for B:

$$17^2 = 10^2 + 14^2 - 2(10)(14)\cos B$$

$$289 = 296 - 280\cos B$$

$$-7 = -280\cos B$$

$$\frac{7}{280} = \cos B$$

$$B = \cos^{-1}\left(\frac{7}{280}\right) \approx 89°.$$

Finally, we can find A because the sum of the angles of a triangle is always $180°$:

$$A = 180° - 89° - 55° = 36°.$$

Therefore, the solved triangle is:

$$\angle A \approx 36°, \quad \angle B \approx 89°, \quad \angle C \approx 55°;$$
$$a = 10, \quad b = 17, \quad c = 14.$$

Solution to practice problem 5: *Solve the triangle: $a = 6$, $b = 8$, $c = 12$.*
First, let's draw a triangle and put in the information that we are given:

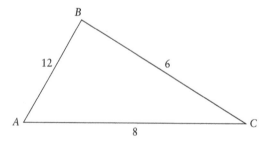

We are given *SSS*, so we will use the Law of Cosines. Let's solve for C:

$$12^2 = 6^2 + 8^2 - 2(6)(8)\cos C$$

$$144 = 100 - 96\cos C$$

UNIT FOURTEEN: The Law of Cosines and Area Formulas

$$44 = -96\cos C$$

$$-\frac{44}{96} = \cos C$$

$$C = \cos^{-1}\left(-\frac{44}{96}\right) \approx 117°.$$

Now let's solve for B:

$$8^2 = 6^2 + 12^2 - 2(6)(12)\cos B$$

$$64 = 180 - 144\cos B$$

$$-116 = -144\cos B$$

$$\frac{116}{144} = \cos B$$

$$B = \cos^{-1}\left(\frac{116}{144}\right) \approx 36°.$$

Finally, we can find A because the sum of the angles of a triangle is always $180°$:

$$A = 180° - 36° - 117° = 27°.$$

Therefore, the solved triangle is:

$$\angle A \approx 27°,\ \angle B \approx 36°,\ \angle C \approx 117°;$$

$$a = 10,\ b = 17,\ c = 14.$$

Solution to practice problem 6: *Find the area of the triangle:* $\angle B = 58°$, $a = 27\,\text{cm},\ c = 19$ cm.

First, let's draw a triangle and put in the information that we are given:

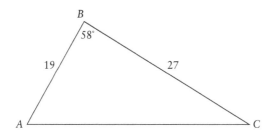

We are given *SAS*, so we can use the formula to find the area, in this case, Area = $\frac{1}{2}ac\sin B$: Area = $\frac{1}{2}(27)(19)\sin 58° \approx 217.52$ cm².

Solution to practice problem 7: *Find the area of the triangle:* $\angle A = 101°$, $b = 10$ ft, $c = 20$ ft.

First, let's draw a triangle and put in the information that we are given:

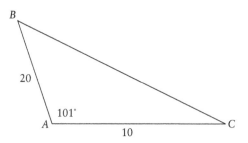

We are given *SAS*, so we can use the formula to find the area, in this case, Area = $\frac{1}{2}bc\sin A$: Area = $\frac{1}{2}(10)(20)\sin 101° \approx 98.16$ ft².

Solution to practice problem 8: *A parallelogram has sides of length 15 in. and 19 in. If the angle between the sides is 88°, find the area of the parallelogram.*

First, let's draw a parallelogram and put in the information that we are given:

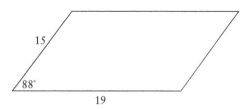

Find the area of a parallelogram by simply multiplying the triangle formula by 2:

$$\text{Area} = ab\sin C.$$

The area is: Area = $(15)(19)\sin 88° = 284.83$ in.²

Solution to practice problem 9: *A surveyor stands some distance away from a pond. The surveyor measures the distance to one edge of the pond as 105 ft. She then turns through 50° and measures the distance to the other edge of the pond as 122 ft. Approximately how wide is the pond?*

UNIT FOURTEEN: The Law of Cosines and Area Formulas 167

First, let's make a sketch and put in the information that we are given:

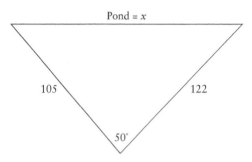

Notice that we are given *SAS*, so we will use the Law of Cosines. Let's solve for the width of the pond, which we have labeled x:

$$x^2 = 105^2 + 122^2 - 2(105)(122)\cos 50°$$

$$x^2 = 25909 - 25620\cos 50°$$

$$x^2 \approx 9440.78141$$

$$x \approx 97.16 \text{ ft.}$$

Solution to practice problem 10: *Find the area of a regular octagon (8 sides) inscribed in a circle of radius 8 cm.*

First, let's make a sketch and put in the information that we are given:

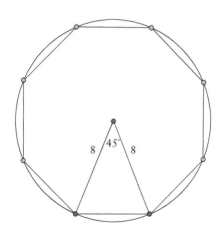

Notice that we can divide the octagon into eight congruent triangles. Each central angle will be $\dfrac{360°}{8} = 45°$, and the radii will form congruent sides of length 8.

We can use the area formula to find the area of one of the triangles and then multiply by 8 to find the area of the octagon:

$$A = \frac{1}{2}(8)(8)\sin 45° \approx 22.6274$$

$$(8)(22.6274) \approx 181.02 \text{ cm}^2.$$